Embedded WiSeNts

IST-004400

Embedded WiSeNts Research Roadmap

Pedro José Marrón, Daniel Minder
and the Embedded WiSeNts Consortium

November 2006

Editors:

Pedro José Marrón
Universität Stuttgart
IPVS, Distributed Systems Group
Universitätsstr. 38
D-70569 Stuttgart
Germany
Tel.: +49-711-7816-223
Fax: +49-711-7816-424
email: pedro.marron@informatik.uni-stuttgart.de

Daniel Minder
Universität Stuttgart
IPVS, Distributed Systems Group
Universitätsstr. 38
D-70569 Stuttgart
Germany

The corresponding editor of this document is Pedro José Marrón

Bibliografische Information der Deutschen Nationalbibliothek

Die Deutsche Nationalbibliothek verzeichnet diese Publikation in der Deutschen Nationalbibliografie; detaillierte bibliografische Daten sind im Internet über http://dnb.d-nb.de abrufbar.

ISBN 3-8325-1424-4

Logos Verlag Berlin
Comeniushof, Gubener Str. 47,
10243 Berlin
Tel.: +49 030 42 85 10 90
Fax: +49 030 42 85 10 92
INTERNET: http://www.logos-verlag.de

Embedded WiSeNts Consortium

The `Embedded WiSeNts` project was created in September 2004 as a Coordination Action funded by the European Commission until December 2006 under IST/FP6 (IST-004400) to create a series of studies on the state of the art of Cooperating Objects and to devise a research roadmap that could be used for the definition of future research programs within the European Commission.

The Embedded WiSeNts Consortium is composed of the following institutions and partners:

- Technische Universität Berlin, Germany (TUB), Coordinator
  - Adam Wolisz
  - Vlado Handziski
  - Irina Piens
- University of Cambridge, UK (UCAM)
  - Andy Hopper
  - George Colouris
  - Marcelo Pias
- University of Copenhagen, Denmark (KU)
  - Philippe Bonnet
  - Eric Jul
- Swedish Institute of Computer Science, Sweden (SICS)
  - Laura Marie Feeney
  - Thiemo Voigt
- University Twente, Netherlands (UT)
  - Paul Havinga

  - Nirvana Meratnia

- Yeditepe University, Turkey (YTU)
  - Sebnem Baydere
  - Onur Ergin
- Consorzio Interuniversitario Nazionale per l'Informatica, Italy (CINI)
  - Chiara Petroli
- University of Padua, Italy (DEI)
  - Michele Zorzi
  - Andrea Zanella
- Swiss Federal Institute of Technology Zurich, Switzerland (ETHZ)
  - Friedemann Mattern
  - Kay Römer
  - Silvia Santini
- Asociación de Investigación y Cooperación Industrial de Andalucía, Spain (AICIA)
  - Aníbal Ollero
  - Iván Maza
- Institut National de Recherche en Informatique et en Automatique, France (INRIA)
  - Michel Banâtre
  - Paul Couderc
- Universität Stuttgart, Germany (USTUTT)
  - Kurt Rothermel
  - Pedro José Marrón
  - Daniel Minder

Please, see `http://www.embedded-wisents.org/` for more information.

Supported by:

SIEMENS VDO

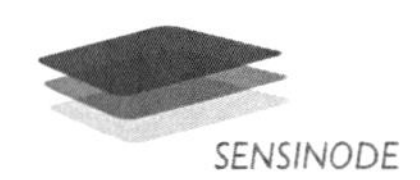

THALES

TURKCELL

SAP RESEARCH

Xanadu Wireless

Acknowledgements

Our thanks go to all our industrial partners, research institutions and colleagues that have taken the time to read this document and give us feedback and very valuable input. This document could not have been finished in its current form without their contributions.

Supporting Institutions

- Ambient Systems B.V.
- Appear Networks
- Barsan Global Logistics
- Communication Research Labs Sweden AB
- CoBIs – Collaborative Business Items
- Continental
- Deutsche Telekom Laboratories
- E-Senza Technologies
- Genetlab A. Ş.
- GoodFood Project
- Gordion
- Hitachi
- Infineon
- LogicaCMG
- Microsoft, Innovation Centre Europe

- Philips
- SAP Research
- STMicroelectronics
- Schneider Electric
- Selex, Sensors and Airborne Systems
- Sensinode Ltd.
- Siemens Automotive
- SmartMote
- Telecom Italia
- Thales
- Turkcell
- Twente Institute for Wireless and Mobile Communications
- Ubisense
- Vodera Ltd.
- Xanadu Wireless

We acknowledge the individual contributions from:

- Mustafa Aykut, Turkcell
- Heribert Baldus, Philips Research Europe
- Sadik Baydere, Business Development Director, Barsan Global Logistics
- Fabio L. Bellifemine, Telecom Italia
- Jan Benders, Senior consultant embedded/RF, LogicaCMG
- Hartmut Benz, Twente Institute for Wireless and Mobile Communications
- Serge Boverie, Siemens Automotive

- Götz Brasche, European Microsoft Innovation Center
- Pierre Chevillat, IBM Research
- Francesco Chiti, University of Florence
- Giuliano Crevatin, RT RadioTrevisan Elettronica Industriale
- Siebren de Vries, CEO, Chess Embedded Technology B.V.
- Alisa Devlic, Research engineer, Appear Networks
- S.O. Dulman, Chief Software Architect, Ambient Systems B.V.
- Stephen Edwards, Technical Director, Whistonbrook Technologies Ltd.
- Richard Egan, Chief Engineer, Thales Research & Technology
- Kemal Egemen, Genetlab A.Ş.
- Stephan J. Engberg, Priway ApS
- Romano Fantacci, University of Florence
- Tapio Frantti, VTT Technical Research Centre of Finland
- Sylvain Gatepaille, Information Processing Competence Center, EADS Defence and Security Systems SA
- Hakan Gültekin, Software Engineer, Barsan Global Logistics
- Thomas Herndl, Infineon Technologies
- Steve Hodges, Hardware Design Engineer, Microsoft Research
- Vania Joloboff, Silicomp AQL
- Björn Karlsson, CEO, Communication Research Labs Sweden AB
- Stamatis Karnouskos, Senior Researcher, SAP AG
- Wolfgang Kellerer, DoCoMo Communications Laboratories Europe GmbH
- Enno Kelling, Manager Electronics Advanced Engineering, Continental Automotive Systems
- Harry Kip, R&D manager, Nedap N.V.

- Costis Koumpis, Vodera Ltd.
- Sudha Krishnamurthy, Deutsche Telekom AG Laboratories
- Markus Krüger, TTI GmbH - TGU Smartmote
- Massimiliano Lenardi, Senior Research Engineer, HITACHI Europe
- Thomas Lentsch, Infineon Technologies
- Jason Lepley, Scientific Consultant, SELEX Sensors and Airborne Systems
- Cees Links, CEO, Xanadu Wireless
- Anders Lundström, Director of R&D, Communication Research Labs Sweden AB
- Gianfranco Manes, University of Florence
- Diego Melpignano, Advanced System Technology - STMicroelectronics
- Matthias Neugebauer, European Microsoft Innovation Center
- Christoph Niedermeier, Senior Principal Engineer, Siemens Corporate Technology
- Ilker Oyman, R&D Director and Software Manager, Gordion Bilgi Hizmet Ltd.
- Davide Di Palma, University of Florence
- Denis Rouffet, Alcatel
- Kilian Schlöder, Deutsche Telekom AG Laboratories
- Ronald Schoop, Vice-President HUB, Schneider Electric
- Dr. Schott, IBM Research
- James W. Scott, Intel Research Cambridge
- Amit Shah, CEO, E-Senza Technologies GmbH
- Zach Shelby, Managing Director, Sensinode Ltd.
- Frank Siegemund, European Microsoft Innovation Center
- Rudolf Sollacher, Principal Research Scientist, Siemens AG
- Pete Steggles, Chief Software Architect, Ubisense

- Guido Stromberg, Infineon Technologies
- Peter van der Stok, Philips Research Laboratories
- Laura Vanzago, STMicroelectronics
- Jens Wukasch, T-Systems

Table of Contents

List of Figures

List of Tables

1 EXECUTIVE SUMMARY

According to a market study performed by ON World Inc. on Wireless Sensor Networks called "Wireless Sensor Networks – Growing Markets, Accelerating Demands" from July 2005, 127 million wireless sensor network nodes are expected to be deployed in 2010 and the growth of this market later on is expected to increase in certain application domains.

Wireless Sensor Networks are a canonical example of a wider field dealing with Cooperating Objects that attempts to create the necessary technologies to make Weiser's vision of the disappearing computer a reality. Cooperating Objects are, in the most general case, small computing devices equipped with wireless communication capabilities that are able to cooperate and organise themselves autonomously into networks to achieve a common task.

The book you have in your hands contains information about the research roadmap envisioned for Cooperating Objects by the `Embedded WiSeNts` consortium and its associated industrial and academic partners. The `Embedded WiSeNts` project was created in September 2004 as a Coordination Action funded by the European Commission under IST/FP6 (IST-004400) to create a series of studies on the state of the art of Cooperating Objects and to devise a research roadmap that could be used for the definition of future research programs within the European Commission. This book is the final result of these two years of work.

1.1 Definition of Cooperating Objects

A number of different system concepts have become apparent in the broader context of embedded systems over the past couple of years. First, there is the classic concept of **embedded systems** as mainly a control system for some physical process (machinery, automobiles, etc.). More recently, the notion of pervasive and **ubiquitous computing** started to evolve, where objects of everyday use can be endowed with some form of computational capacity, and perhaps with some simple sensing and communication facilities. However, most recently, the idea of **wireless sensor networks** has started to appear, where entities that sense their environment not only operate individually, but collaborate together using ad-hoc network technologies to achieve a well-defined purpose of supervision/monitoring of some area, some particular process, etc.

We claim that these three types of systems that act and react on their environment are actually quite diverse, novel systems that, on the one hand, share some principal commonalities and, on the other hand, have some different aspects that complement each other to form a coherent group of objects that cooperate with each other to interact with their environment. In particular, important notions such as control, heterogeneity, wireless communication, dynamics/ad-hoc nature, and cost are present to various degrees in each of these types of systems.

The conception of a future-proof system would have to combine the strong points of all three system concepts at least in the following functional aspects:

- Support the control of physical processes in a similar way embedded systems are able to do today;
- have as good support for device heterogeneity and spontaneity of usage as pervasive and ubiquitous computing approaches have today;
- be as cost efficient and versatile in terms of the use of wireless technology as wireless sensor networks are.

For these reasons, these new systems consist of individual entities or objects that jointly strive to reach a common goal, which involves sensing or the controlling of devices, and are dynamically and loosely federated for cooperation. All of this, while making sure resources are used optimally.

1.2 State of the Art in Cooperating Object Research

In this document, we present a summarised version of the contents of four studies conducted in each of the following areas: applications and application scenarios, paradigms for algorithms and interactions, vertical system functions, and system architecture and programming models. The full-fledged version of these studies can be found in the `Embedded WiSeNts` website[1].

Applications and Application Scenarios: In this study we have provided a characterisation of Cooperating Object applications and classified currently available projects into this framework. The following application domains and projects associated with them have been investigated:

- Control and Automation
- Home and Office

[1] `http://www.embedded-wisents.org/`

- Logistics
- Transportation
- Environmental Monitoring for Emergency Services
- Healthcare
- Security and Surveillance
- Tourism
- Education and Training

The projects listed in this categories have been investigated from a research point of view and show in most cases integrated solutions that work well for specific application domains.

Paradigms for Algorithms and Interactions: In this study we have discussed four thematic areas where algorithms and interactions can be mapped to: Wireless Sensor Networks for Environmental Monitoring, Wireless Sensor Networks with Mobile Nodes, Autonomous Robotic Teams, and Intervehicle Networks. Then, based on the characteristics and requirements defined in the previous study, we have provided a characterisation of the following algorithms and interactions:

- MAC Algorithms
- Routing Algorithms
- Localisation Algorithms
- Data Processing
- Navigation Algorithms

These algorithms, while being interesting from the point of view of research, also constitute a list of the kinds of services that can be found as part of the integrated solutions described in the previous study.

Vertical System Functions: In this study we have taken care of vertical system functions, that is, functionality that, due to their nature needs to be taken care of throughout the entire application, an not in only one layer of the more traditional layered approach. The types of vertical system functions discussed in this study are the following:

- Context and Location Management
- Data Consistency
- Communication Functionality
- Security, Privacy and Trust

- Distributed Storage and Data Search
- Data Aggregation
- Resource Management
- Time Synchronisation

As for the algorithms described in the previous study, these vertical system functions are found in the integrated solutions addressed in the first study and show the complexity of the field and the fact that classic architectures need to be redesigned with cross-layer optimizations and vertical interactions in mind.

System Architecture and Programming Models: In this study we have classified currently available architectural models and middleware approaches that can be used to abstract the complexity of Cooperating Object technology. The following topics have been discussed in detail:

- Virtualisation
- Operating Systems
- Adaptive Middleware Approaches
- Data Distribution and Addressing
- Self-organisation and Regulation
- Global Information Spaces
- Support for Control Synchronisation

In order to validate some of the results detailed above, the members of the `Embedded WiSeNts` consortium have performed a survey among the participants of the EU Workshop "From RFID to the Internet of Things" held in Brussels on March 6th/7th 2006. In it, we prepared a series of questions that were posed to volunteers that answered in an anonymous way. We obtained answers from 51 people working in both, university and industrial research and product development in the industry.

Over 80% of all participants estimate that the breakthrough for the technology will happen in the next 5 to 10 years and it is safe to assume that they assume that most technological inhibitors and research issues will be solved by then.

As expected, there is a certain number of optimists that believe Cooperating Object technology is almost ready for prime-time, and also a slightly larger number of pessimists (or realists?) that do not expect Cooperating Objects to be widely used in the industry until at least 15 years from now.

1.3 Research Gaps and Timeline

Looking at the state of the art described in the previous sections, we have been able to identify a series of gaps that we believe to be interesting from the point of view of research for the coming years. These gaps have been grouped into 5 different categories as follows:

Hardware: These are gaps that have to do with the development of the devices that physically constitute networks of Cooperating Objects. The gaps that fall into this category are: *Sensor Calibration*, *Power Efficiency*, *Energy Harvesting*, *New Sensor and Low-Cost Devices* and *Miniaturisation*.

Algorithms: These are gaps that deal with functional properties of Cooperating Objects, that is, specific protocols, types of procedures, etc. The gaps that fall into this category are: *Localisation*, *Context-aware MAC and Routing*, *Clustering Techniques*, *Data Storage and Search* and *Motion Planning*.

Non-functional Properties: These are gaps that deal with Quality-of-Service-type characteristics. Properties that do not affect the functionality of the network, but its quality. The gaps that fall into this category are: *Multiple Sinks*, *Scalability*, *Quality of Service*, *Robustness*, *Mobility*, *Security*, *Heterogeneity* and *Real-time*.

Systems: These are gaps that have to do with the specific architecture or support for the rest of the system. Normally, systems work at the individual Cooperating Object level, but have to provide support for networking. The gaps that fall into this category are: *Adaptive Systems*, *Operating Systems*, *Programming Models* and *System Integration*.

Others: This category collects the gaps that do not fit anywhere else or that might be hard to classify within other categories because they do not really fall into the computer science umbrella. The ones we have selected for description here are the following: *Modeling and Analysis*, *Experimentation* and *Social Issues*.

In general, we assume the largest number of breakthroughs in the areas mentioned above to happen in the middle-term, that is, between 5 and 10 years, which agrees with the estimation obtained from our survey regarding the point in time where Cooperating Objects will start to be used widely in the industry.

Most of these are already being worked on in some form or another. Only two of the gaps are expected to start 5 years from now: *Real-time* and *Social Issues*. The first one because of the nature of the problem and the second one because social issues will only arise as soon as the early adopters (especially from industry) start introducing Cooperating Objects more aggressively in our everyday lives. In general, gaps that are not being investigated yet or that need investigation throughout the predicted period seem to be the most promising lines of research.

1.4 Potential Roadblocks

Working with members of the industry and with the participants of the Embedded WiSeNts survey, it seems clear that, although the opinion of all experts indicates that Cooperating Object technology has clear chances of success, there is always the possibility of failure if certain issues are not solved properly or in a timely manner.

Here are the potential roadblocks that have appeared during these conversations and interactions:

No Clear Business Models: One of the main potential roadblocks for the adoption of any kind of technology is the lack of a business model that supports it. For the case of Cooperating Objects, it is probably too early to determine whether or not strong enough business models will appear and, as far as early adopters are concerned, there are enough examples of companies that make their living nowadays selling technology that can and will be used in this field. However, it might be necessary to work tightly with end-users to identify the real needs and, therefore, business models with high potential.

Lack of Standards: With the creation of a new field, it is obvious that early adopters need to provide a pragmatic solution in order to "show something that works", but after a certain time, the industry needs to come together and agree on a common ground for future developments. There are already some attempts to standardise ZigBee and UWB, which will probably play an important role as communication protocols that bring together networks of Cooperating Objects. However, this is just the beginning and further actions need to be taken in order for Cooperating Object technology to take off.

Confidence in Technology: For more sophisticated applications, currently available Cooperating Object technology is, for the most part, not able to deliver the desired characteristics, such as lifetime or robustness. Therefore, the immaturity of current solutions in individual fields hinders the adoption of Cooperating Objects in more general application domains. However, in some cases this reluctance is based on prejudices, e.g. against wireless communication. It is hard to convince people that a new technology which is generally considered as more error-prone can deliver almost the same quality of service as the old, wired technology when designed carefully.

Social Issues: Even if the technological issues are solved and the industry is able to pull together a set of standards that supports Cooperating Object technology, the end-users are still the ones that decide whether or not they will want to make use of this technology. The main question is whether or not the vast majority of people is willing to accept tiny computing devices "interfering" with their lives. People are not willing to have "big brother" watching them unless they see a benefit to it. In general, people are reluctant to provide private information that might give an insight on their daily activities or habits and, therefore, if Cooperating Objects can be misused for this purpose, finding a killer-application might take longer than expected, if at all. An increase in the awareness of security and privacy issues is surely needed for the proper understanding of the capabilities of Cooperating Objects, so that the end-users can put this new technology into perspective and not feel threatened.

1.5 Recommendations

Several issues have been identified by the research community, the industry and end-users as important work items. We have identified three different types of actions: actions related to research, actions related to educational activities and other actions.

Research Actions: Regarding research, there are five major topics that should be addressed and that summarise some of the major gaps listed above:

- Research on algorithms, such as data search and storage, aggregation, consistency and the integration of Cooperating Objects into existing software.
- Research on hardware, such as new small, cheap components, the miniaturisation of existing ones and energy harvesting techniques for their powering.
- Research on non-functional properties, such as Quality of Service, power consumption, scalability, mobility and security.
- Research on adaptive systems that are easy to use, easy to combine with existing software, adapt to changes in the environment and are fully integrated in operating system and programming abstraction solutions.
- Research on system integration and combination of heterogeneous techniques into one working system.
- Research on social issues of Cooperating Objects in order to improve on the understanding of such systems both for research and for possible commercialisation.

Research on all topics should be application-driven to ensure the convergence of research and applications. Real-world experiments should be carried out in order to understand and solve the underlying scientific challenges of providing practical and efficient solutions to real world problems.

Educational Actions: There is a need for education of the industry and of end-users in order to promote the new technology and to allow for potential users and customers to understand the benefits and risks associated with the use of Cooperating Objects. At the same time, companies that would like to enter this market need to be taught about the potential and the power of this new technology so that they can properly address the needs of their customers.

Other Actions: The most prominent action that needs support from all major players is the process of standardising the hardware and software available for Cooperating Objects.

There will also be a need to support the creation of regulations and legislative actions that create a legal framework that supports the correct use of Cooperating Objects and hinders possible misuses of this technology.

2 PURPOSE OF THIS DOCUMENT

The document you have in your hands presents the vision of the Embedded WiSeNts consortium and its associated industrial partners regarding the future development of research in the field of Cooperating Objects. This vision is presented in the form of a technology roadmap and is the result of the compilation of several factors:

- The individual **expertise** and **practical experiences** of each of the partners involved in the project;
- the analysis of the current technologies and **current trends** that show future research directions;
- a **market analysis** of Cooperating Objects performed with the input from industrial partners and other research institutes;
- **visionary applications** obtained from a wider audience;
- and the **identification of gaps** and research agendas in the different areas that compose the field of Cooperating Objects.

Given the research oriented background of all partners involved in the project and the fact that the field of Cooperating Objects is advancing rapidly, this document does not pretend to present a roadmap that can be used "as is" to drive development for any industrial sector involved in this field. It can, however, be used as further input for development and innovation departments that would like to benefit from information about the possible direction and the timeframe for Cooperating Object research at the European level.

The final part of the document contains a series of recommendations that can be used by financing institutions and organisations in order to drive research in the direction shown in this document.

2.1 Intended Audience

The Embedded WiSeNts Research Roadmap has been written with three different audiences in mind, as follows:

Researchers: That work or intend to work in the field of Cooperating Objects and would like to understand the current state of the art, current trends and possible gaps for future research.

Industry: That would like to understand the current state of the art and possible market developments to be used as an additional source of information for the definition of specific strategies and business opportunities related to Cooperating Objects.

European Commission: To achieve a better understanding of the field of Cooperating Objects and its potential as a topic that can be included in upcoming framework programmes or other financing instruments within the EU.

Depending on the interest of the reader and its adhesion to one or more of the audiences described above, the reader should select the chapters and sections that most fit his/her interests. The following section identifies the different parts of the roadmap, how they fit together and the kind of information that can be expected from them.

2.2 Methodology and Structure

Figure 2.1 gives an overview of the different parts of this document and of the relationship between its chapters. The boxes represent main topics discussed in the roadmap that can be usually found in different chapters or sections within a chapter. The arrows represent inputs and preliminary information that has been used in the process of writing the corresponding sections.

The figure has been colour coded using the following convention: blue boxes represent chapters that have been obtained from other work packages during the execution of the `Embedded WiSeNts` project. The information contained in these chapters is a summary with the relevant data from longer documents and reports that can be obtained from the project website [12]. Gray boxes represent documents that have been obtained from external sources to the project, such as companies that provide market information used in this roadmap to identify market trends and predictions. Green boxes show intermediate results derived from other chapters and additional documents. These intermediate results are in themselves interesting for certain audiences that would like to understand the process used to get to the final results. Finally, the yellow boxes represent the actual output of this roadmap (final results) in terms of recommendations for further work for the three types of audiences described above.

Following the graphical representation in Figure 2.1, the structure of the document and the methodology used to write it is as follows. Chapter 4 contain a summarised description of the contents of four studies developed as part of the `Embedded WiSeNts` project that define the state of the art in terms of current applications and application scenarios,

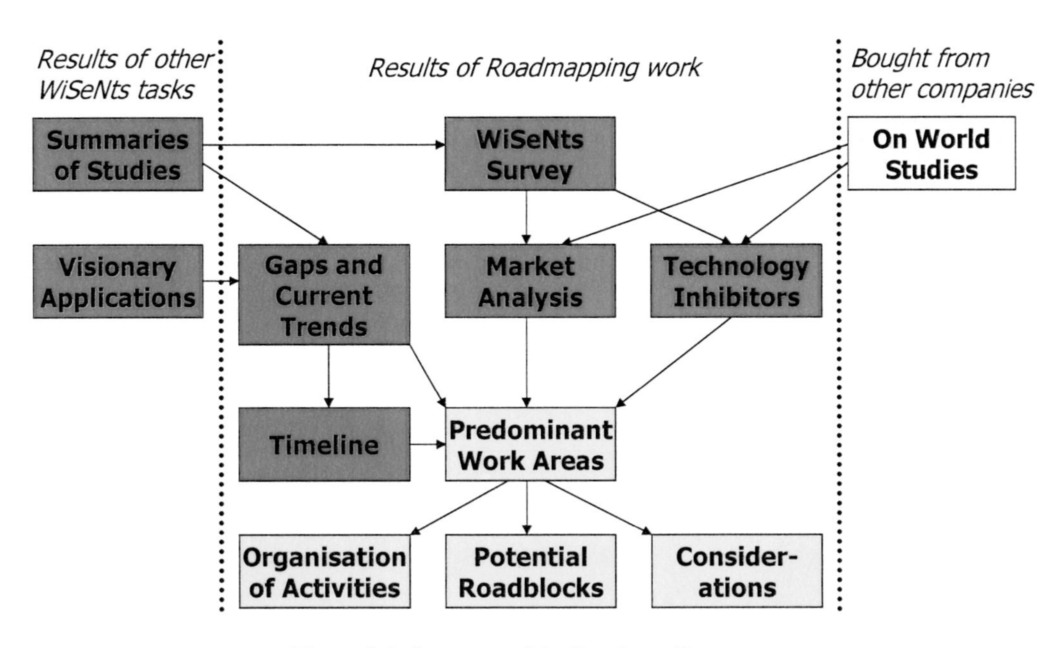

Figure 2.1: Structure of the Roadmap Document

paradigms for algorithms and interactions, vertical system functions and system architecture and programming models. These and the summary of the visionary applications in chapter 5 form the main input for the Gaps and Current Trends section, that can be read in section 7.1. A mapping of these gaps and current trends into a timeline that extends from today up to 15 years in the future can be found in section 7.2.

Following a parallel track, the results of the studies have been evaluated and validated in a survey among the participants of the "From RFID to the Internet of Things" workshop organised by the EU. Using the information obtained there and data from the market analysis we have bought from ON World Inc., we were able to provide a market analysis study in chapter 6. These same inputs were also used to determine the most prominent inhibitors for Cooperating Objects described in section 7.4.

These four intermediate results (gaps and current trends, timeline, market analysis and inhibitors) were used to determine the predominant work areas in section 8.1, that served as a driving force and input for the final results of the roadmap, namely, the organisation of activities (section 8.2), potential roadblocks or major inhibitors that hinder the acceptance of Cooperating Object technologies in society (section 8.3), and the final considerations for research programs in section 8.4.

Using Figure 2.1, the reader can decide what chapters and sections of the roadmap to read and whether or not only the final results and summaries are of interest, or also the process towards these results are worth the reader's time.

3 INTRODUCTION TO COOPERATING OBJECTS

A number of different system concepts have become apparent in the broader context of embedded systems over the past couple of years. First, there is the classic concept of **embedded systems** as mainly a control system for some physical process (machinery, automobiles, etc.). More recently, the notion of pervasive and **ubiquitous computing** started to evolve, where objects of everyday use can be endowed with some form of computational capacity, and perhaps with some simple sensing and communication facilities. However, most recently, the idea of **wireless sensor networks** has started to appear, where entities that sense their environment not only operate individually, but collaborate together using ad-hoc network technologies to achieve a well-defined purpose of supervision of some area, some particular process, etc.

We claim that these three types of systems that act and react on their environment are actually quite diverse, novel systems that, on the one hand, share some principal commonalities and, on the other hand, have some different aspects that complement each other to form a coherent group of objects that cooperate with each other to interact with their environment. In particular, important notions such as control, heterogeneity, wireless communication, dynamics/ad-hoc nature, and cost are present to various degrees in each of these types of systems.

The conception of a future-proof system would have to combine the strong points of all three system concepts at least in the following functional aspects:

- Support the control of physical processes in a similar way embedded systems are able to do today;
- have as good support for device heterogeneity and spontaneity of usage as pervasive and ubiquitous computing approaches have today;
- be as cost efficient and versatile in terms of the use of wireless technology as wireless sensor networks are.

For these reasons, these new systems consist of individual entities or objects that jointly strive to reach a common goal, which involves sensing or the controlling of devices, and are dynamically and loosely federated for cooperation. All of this, while making sure resources are used optimally.

A possible term for such a new system conception, born out of the combination traditional embedded systems, pervasive/ubiquitous computing, and wireless sensor networks are **"cooperating objects & pervasive control"**, that stress the point that a participating object does not need to be a single physical entity, but can very well be a composed object in and of itself – a wireless sensor network would be a typical example, making them an important, if not a canonical class of cooperating objects.

Moreover, this notion or paradigm of Cooperating Objects is even stronger as it carries over to the internal structure of such a wireless sensor network as well – in fact, they can be regarded as consisting of cooperating objects themselves, highlighting the diversity of cooperating patterns admissible under this general paradigm. Also, pointing to the importance of complementing the vision of pervasive computing with that of **pervasive control** will be essential.

3.1 Definition

Following the concepts we have just discussed, let us now define what a Cooperating Object is. In the abstract sense, a Cooperating Object is a single entity or a collection of entities consisting of:

- *Sensors*,
- *controllers (information processors)*,
- *actuators* or
- *cooperating objects*

that communicate with each other and are able to achieve, more or less autonomously, a common goal.

More precisely, *sensors* are devices that act as inputs to the Cooperating Object and are able to gather and retrieve information either from other Cooperating Objects or from the environment.

Controllers are devices that act as data or information processors and, obviously, must interact with *sensors* and *actuators* in order to be able to interact with their environment. Furthermore, *controllers* are equipped with some kind of storage device that allows them to perform their tasks. The amount of "effort" devoted by a particular controller to either information processing or storage tasks is determined on an individual basis. This way, the sensor network might be composed of controllers that mostly provide information processing capabilities, whereas others might be specialised in the storage of data.

Finally, *actuators* are devices that act as output producers and are able to interact and modify their environment.

It seems clear that if *sensors*, *controllers* and *actuators* need to interact with each other in a distributed environment, all of them need to be equipped with communication capabilities. These might of course be based on wired or wireless technology.

The inclusion of other *cooperating objects* as part of Cooperating Object itself indicates that these objects can combine their *sensors*, *controllers* and *actuators* in a hierarchical way and are, therefore, able to create arbitrarily complex structures.

As a concrete example, imagine that a Cooperating Object is used to collect temperature gradients of flammable liquid within an industrial plant. When the gradient achieves certain predefined thresholds, safety pipe valves must be opened to minimise the risks of an explosion. In this scenario, we have two Cooperating Objects: one that continuously measures temperatures and another that actuates in the environment by manipulating valves. The first one is an example of a classical sensor network, whereas the second would be traditionally described as an "actuators and controllers network".

The generality of this model allows us to seamlessly include different fields like sensor networks, pervasive computing, embedded systems, etc. However, depending on the way we look at the algorithms and systems developed for Cooperating Objects, we can define **data-centric** and **service-centric** approaches.

3.2 Data-centric Approaches

The field of wireless sensor network research is the canonical example of **data-centric** approaches. In this field, the efficient management of *data* is in the core of all published algorithms. Additionally, some other characteristics are relevant for sensor networks, such as:

- Minimal user interaction: Given a query from the user, a sensor network should be able to autonomously and automatically figure out the most efficient way to provide an answer to the user.
- Resource-limitation: Sensors have usually limited resources in terms of energy, capacity, sensing capabilities, etc.
- Ad-hoc organisation: Most sensor networks expect their nodes to be able to communicate with each other without the use of any kind of infrastructure.
- Wireless communication: As a consequence of the ad-hoc nature of sensor networks, communication is usually performed using wireless technology.

In general, **data-centric** approaches are chosen in environments where the naming of data and the use of data types within the network play a more important role than the specific node that might be responsible for its processing. Therefore, there is a dissociation of data and network node which can be used to dynamically select the appropriate location where data processing is performed without it affecting the expectations from the user. **Data-centric** approaches are best suited for database-like operations like aggregation and data dissemination.

In the literature, there are two different kinds of **data-centric** processing techniques. The first one uses the query/response (or request/reply) paradigm, so that the network of cooperating objects only sends responses to specific queries issued by the user. In the second technique, queries are "stored" in the network and are provided with an associated lifetime. During their lifetime, each sensor is responsible for the processing of the stored (or continuous) query and sends messages to the issuer of the query (also called sink) whenever the condition specified in the query is met. Therefore, both pull-based and push-based approaches can be used in data-centric environments.

Although the absolute position of nodes within the network do not play an important role from the perspective of the user (or the issuer of the query), good topology management techniques need to be used in order to maximise the lifetime of individual sensors. It is crucial to know where the neighbours are and what kinds of roles they play in the network in order to optimise the processing of queries.

Finally, real-time is usually not a concern in the literature since in most cases, real-time and energy optimisation are contradicting goals. In most cases, the use of energy-efficient algorithms is favoured over real-time constraints, so that the use of data-centric approaches in real-time environments has not been thoroughly studied.

3.3 Service-centric Approaches

In contrast, **service-centric** approaches are mostly concerned with the definition of the interface or *API* in order to provide functionality for the user. Depending on the specific fields there are other additional characteristics that need to be mentioned. For example, in the field of pervasive computing, the miniaturisation of devices as well as resource-limitation play an important role, whereas in classic client-server architectures no such restrictions apply.

In such environments, the transport mechanisms are hidden from the user applications (such as in traditional networked environments), but a certain cooperation among the nodes in the network allows for the processing of data. The difference to **data-centric** approaches lies in the kind of programming techniques needed to interact with the network. In a **service-centric** environment, the application developer is supposed to have and use a clear specification of services offered by the network.

Also in these types of environments is the use of pull-based and push-based approaches widespread. The use of traditional APIs would cover the case of pull-based interactions, whereas a publish/subscribe mechanism would provide the necessary APIs needed for a system to interact with the user in a push-based fashion. The specific location of a service might be transparent to the user, but it is usually the case that the application developer needs to know where services are located, or needs to be able to contact a location service that knows where services are located.

Finally, the implementation of real-time APIs and real-time constraints has been studied in much more detail since the enforcement of such constraints can be performed through API implementations.

3.4 Enabling Technologies

It seems clear that, nowadays, it is impossible to create or work on technologies that do not rely more or less heavily on the development of other related areas. New developments in these related areas usually go hand-in-hand, and a major breakthrough in one of the enabling technologies can really boost the work that can be performed on the other areas.

This is also true for Cooperating Objects and, as we have seen in the previous sections, Cooperating Objects have emerged as a combination and natural extension of already existing research areas that have been evolving rapidly in the past years.

Therefore, it is worth pointing out more precisely what we think are the major pillars for research in Cooperating Objects, so that the readers can keep them in the back of their heads while reading the following sections.

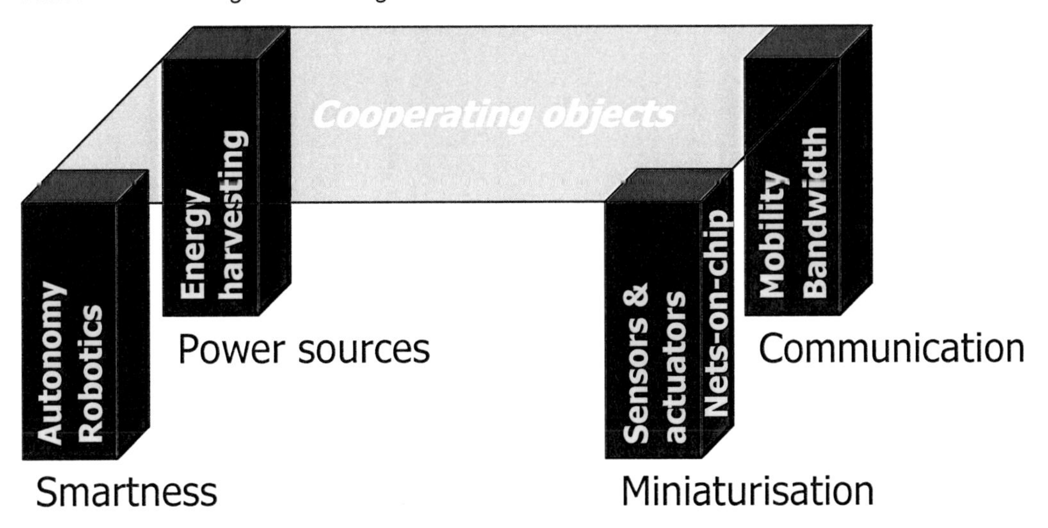

Figure 3.1: Enabling Technologies for Cooperating Objects

Figure 3.1 shows the four pillars that, in our view, support work on Cooperating Objects. These are: Miniaturisation, power sources, communication and smartness. Given a certain predominance of software experts among the members of the `Embedded WiSeNts` consortium, some of these areas, like miniaturisation and power sources, fall outside the expertise of the group. Others, like communication and intelligence, are definitely covered and also appear as part of the gaps and current trends identified in the following chapters.

Let us now describe in more detail what the purpose and relationship of each pillar with Cooperating Object technology.

Miniaturisation: Research on miniaturisation deals with the creation of always smaller sensors, actuators and, in general, devices that can be used to implement a network of Cooperating Objects. If the long-term vision of Mark Weiser of the disappearing computer is to become true, the miniaturisation of devices plays a crucial role by implementing more into less space.

The newest developments in this area even discuss now the possibility of incorporating a whole network of devices into a single chip (nets-on-chip), making them ideal candidates for their incorporation in Cooperating Object research.

Power Sources: Related to the previous enabling technology, research on power sources seems to be one of the major concerns when designing smaller and smaller devices. Current research on batteries and power sources cannot really keep up with microcontroller technology and, when discussing this issue with hardware experts, complain that hardware could be much smaller if they just had a way to power it properly.

Following this suggestion, a considerable amount of effort is being put on energy harvesting techniques that use vibrations, electro-magnetic waves, motion, etc. to power small devices. The good news is that, in most cases, the smaller the devices, the lower the amount of energy is that needs to power it, but current research has not yet found the sweet spot where miniaturisation and battery technology can be scaled down together to the sizes needed for the implementation of Cooperating Object technology.

Communication: Research on communication technologies has received a lot of attention in the past decades and with the success of the Internet, the research community has produced highly efficient algorithms for the transmission of data between computers.

However, the characteristics of Cooperating Objects, especially the fact that devices have to communicate with each other in order to be able to do anything interesting, and the sheer amount of devices that need to communicate, has changed the characteristics and metrics that make communication algorithms efficient. New Quality of Service (QoS) metrics have been identified, such as energy consumption, bandwidth or mobility (shown in Figure 3.1), that make this field not only an enabling technology for Cooperating Objects, since communication is crucial for its operation, but also have also created new research directions that can be followed independently of explicit research on Cooperating Object technology.

Smartness: Also related to communication and QoS is the fact that cooperation needs to happen in an unknown (sometimes even hostile) environment. Therefore, smartness is an enabling technology for Cooperating Objects since communication is definitely necessary but the need for smart behaviour and efficient cooperation is definitely needed from the Cooperating Objects that make up a network. In this area, the autonomy achieved by current robot technology, or the fact that the system and each device needs to sense and adapt itself to its environment, make a certain degree of smartness a required characteristics and, therefore, an enabling technology, for Cooperating Objects.

4 STATE OF THE ART IN COOPERATING OBJECT RESEARCH

Given the breadth of Cooperating Object research depicted in the previous chapter, an in-depth analysis of the state of the art requires the study of several interrelated and complementary topics that span the field of Cooperating Objects. Figure 4.1 provides a graphical representation of the contents of this chapter. Each of the following four sections deal with one of the topics shown in the figure: applications, algorithms and paradigms, vertical functions and architectures.

As indicated in the figure by the use of arrows, applications have an influence on the architecture used by Cooperating Objects that determine the requirements of the system itself. These requirements drive the design and development of paradigms and algorithms used to implement the application. In addition to algorithms located in traditionally layered architectures, the nature of Cooperating Objects and their algorithms, as well as the need for optimisation required by the applications drive the development of vertical functions that span several or all of the layers found in more traditional architectures.

In this chapter, we present a summarised version of the contents of four studies conducted in each of the areas mentioned above in the field of Cooperating Objects. The interested reader can find the full-fledged version of these studies in [12], although the following sections contain not only an in-depth analysis of the current state of the art (as found in the studies), but also the identification of essential open issues (gaps) and current trends and timelines for research in the corresponding area. Also in [12], the report "Critical evaluation of research platforms for wireless sensor networks" can be found which covers hardware platforms, operating systems, programming environments, simulation and emulation environments and testbeds that are not or only partially covered by this roadmap.

More specifically, section 4.1 provides an overview of applications and application scenarios with their typical parameters and requirements. Section 4.2 shows a classification of basic and advanced paradigms used for the design of algorithms and interactions patterns used in Cooperating Objects. Vertical functions and their role regarding optimisation and quality of service are identified in section 4.3, whereas section 4.4 describes and classifies the set of programming models, paradigms and system architectures for Cooperating Objects.

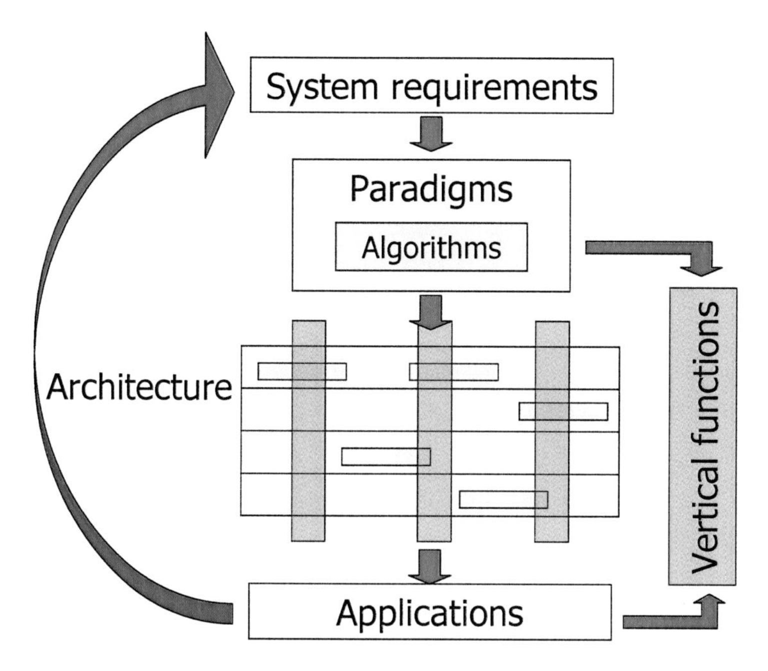

Figure 4.1: Organisation of the Studies

4.1 Applications and Application Scenarios

4.1.1 Introduction

This section provides a summary of Cooperating Objects applications and application scenarios that can be readily understood today. The main objective is to identify relevant state of the art projects and activities in the Cooperating Object domain. For this purpose, both European projects and other projects outside Europe have been surveyed. Although sensor network research were initially driven by military applications such as battle-field surveillance and enemy tracking, several civil applications have emerged in recent years.

The applications and scenarios take into account the state of the art of current service-centric (control applications, pervasive or ubiquitous computing) as well as data-centric approaches (wireless sensor networks). In data-centric approaches efficient management of data is the major concern whereas service-centric approaches are mostly concerned with the definition of the interface or API in order to provide functionality for the user. Hybrid scenarios where service-centric and data-centric technologies must be combined also exist. Hybrid scenarios and applications have the potential to provide valuable clues for the identification of distinguishing and overlapping features of service-and data-centric approaches. Since the history of service-centric approach is older than data-centric one and most service-centric requirements can be satisfied by current networking technologies, this identification is useful for drawing a roadmap of challenging research areas.

The characteristics of Cooperating Object applications are quite different from traditional wireless and wired networks. There are critical factors that influence the architectural design and protocol development of Cooperating Object applications that introduce some stringent constraints. Moreover, the constraints for a Cooperating Object application domain may be quite different from another one. For example, security requirements in health and security applications can be more critical than in a home application scenario. Therefore, a one-size-fits-all approach does not work for Cooperating Objects. Additionally, there is a need for application categorisation in order to be able to identify common requirements.

Cooperating Object applications can be classified in many different ways since each application has common features with others such as mobility, scalability, heterogeneity, robustness, fault tolerance, etc. In this study, application domains that have a social and economic impact in society are used as the basis of classification. These domains are control and automation, home and office, logistics, transportation, environmental monitoring and emergency services, health-care, security and surveillance, tourism, education and training.

The following sections are organised as follows: First, we define general characteristics and requirements of Cooperating Object applications to derive a taxonomy. Then we provide an analysis of common characteristics and application dependent features of each taxonomy, summarising all the characteristics for the application taxonomies studied here in a table for easy comparison. Finally, we identify current trends and conclude this section with a timeline of the studied applications, indicating their last known status. It is obvious that the number of applications and projects presented here do not cover all known applications. For this study, we have restricted our search to projects and application domains whose information is publicly available which, in the end, turned out to be about 40 projects.

4.1.2 Characteristics and Requirements of Cooperating Object Applications

In this section we enumerate and briefly explain the typical architectural trends that best characterise the Cooperating Object applications and system requirements that influence the protocol and algorithm design in the wide sense of the term. These characteristics are used as the basis of analysis of open research issues in different application domains. State of the art research on common characteristics are given in sections 4.2 and 4.3. Also, in [55], a good categorisation of the requirements and characteristics of wireless sensor networks is presented. The following three properties of Cooperating Object applications are general *characteristics* they may possess.

Data Traffic Flow: The amount of data sent across the network determines the traffic characteristics of an application. In one application the data transferred among nodes can be limited to a few bytes for simple measurements whereas heavy audio/video traffic can be the main traffic component in another application. Potentially, wireless sensor network traffic does not follow any known traffic patterns. It is non-stationary and highly correlated because of the event-driven characteristic of the Wireless Sensor Network. When an event is detected, there are sporadic outbursts of high traffic; otherwise, most sensor nodes will remain asleep for long durations. The traffic characterisation of Wireless Sensor Networks is very complex and difficult. All layers of the protocol stack affect the traffic pattern of the network.

Multipath phenomenon, human activities, background noise, node orientation, and interference from other nodes cause to severe changes in traffic pattern of a Wireless Sensor Network. The protocols running on the network layer have significant effects on the traffic pattern of the sensor network as well. For example, the traffic characteristics change if two packets with the same destination are combined into one packet. This is a kind of data aggregation.

Network Topology: In a Cooperating Object application, sensor nodes may directly communicate with an actuator, or a sensor node sends its data to the actuator through several sensors. The first case implies a *single hop* topology whereas the second requires a *multi-hop* sensor network. With current technology, single-hop communication model is easier to establish than multi-hop communication networks. Multi-hop topologies have significant challenges such as routing, support for mobility and scalability, etc. Substantial research is still needed on the requirements of multi-hop setups in real life applications.

Indoor vs. Outdoor: Generally speaking, operating environments for Cooperating Object applications are categorised as *indoor* or *outdoor*. Indoor applications are mostly implemented in home and office environments whereas roadways, railways, and forests may be some examples of outdoor environments for Cooperating Object applications. Most of the factors listed in this section, such as localisation, security and mobility introduce more challenging requirements for outdoor applications.

The other properties we state below, are the general *system requirements* of Cooperating Object applications and, depending on the task they are focuses to solve, might give more or less importance to each of the properties specified below.

Autonomy: Nodes can be remotely controlled or fully unattended and autonomous. In instances of applications of the latter class, nodes make autonomous decisions according to the collected information. Particularly, the applications operating without human intervention and including robots and automated machinery require a high degree of autonomy.

Context Awareness: In Cooperating Object applications, some devices or objects may need to have information about the circumstances under which they operate and can react accordingly. These context aware objects may also try to make assumptions about themselves or the current status of the objects they monitor or control.

Fault Tolerance: The loss of sensor nodes during operation of the network due to limited power capacity and challenging operation environment is intrinsic to Cooperating Object applications. In many outdoor applications, it is very hard or even impossible to change batteries of sensor nodes and, as a result, a Cooperating Object network must be able to sustain its operation even in the presence of failures.

Localisation: There are several Cooperating Object applications for target tracking and event detection, e.g., intrusion, forest fire, etc., that need node and/or target localisation. For this purpose satellite based positioning systems are used. The most popular positioning system is GPS (Global Positioning Systems) and it can be used in applications where scalability and cost per node requirements can be satisfied. However, the cost of equipping every node with a GPS unit cannot be tolerated in many applications. Also, mounting GPS receiver to sensor nodes increases their size and their power consumption which is not wanted. Furthermore, in some environments such as indoor and undersea, GPS does not work. There are also some proposed GPS-free localisation schemes for wireless sensor networks, however it is a significant challenging issue yet.

Mobility: In some applications, all physical components of the underlying system may be static whereas in others, they may contain mobile nodes. Especially applications which can benefit from autonomous robots in the field of action may require special support for mobility. Mobility support for multi hop routing in infrastructure-less networks is still a challenging issue. High mobility requirement of the application also affects the design for other characteristics such as localisation and synchronisation.

Networking Infrastructure: Cooperating Object networks can be either *infrastructured* or *infrastructure-less (ad hoc)*. Even in some applications the data can be collected by some mobile nodes when passing by the source nodes. Having an infrastructure or not mostly depends on the operational area of the Cooperating Object application. For example, some environmental monitoring and surveillance applications established in remote regions require infrastructure-less operation, whereas others may benefit from other wireless and/or wired systems in the environment.

Node Heterogeneity: Most of Cooperating Object applications include different types of nodes that have distinct hardware and software characteristics. For instance, in a precision agriculture application, there may exist various sensor types like biological, chemical, temperature and humidity sensors.

Packaging for Robustness: In many Cooperating Object applications, the low-power, low-bandwidth, tiny sensors will be used in challenging operational environments. For instance, in a desert, sensor nodes must be covered with some housing structure in order to prevent them from high temperature or other harsh desert conditions.

Power Awareness: It is obvious that power consumption is one of the most crucial performance metrics and limiting factors in almost every Cooperating Object application. In order to make a system practical for real world scenarios which require long life-time, efficient power consumption strategies must be developed. Long-lifetime is expected especially from an application established for environmental monitoring. For instance, a Cooperating Object application set up in a remote site such as a desert, must have a long lifetime, since it is not easy to access that environment and to replace the network with a new one. There are lots of researches in the literature focused on developing more power efficient strategies.

Production and Maintenance Cost: Depending on the application type, Cooperating Object applications containing a large number of nodes and aiming at operating for a long time require low production or low maintenance cost. This need can be determined by some characteristics of the applications. For example, for the networks which are expected to stay alive and operate for a long time, low maintenance cost becomes more important, which can be achieved through higher production costs. Also, these cost constraints have a great influence on the capabilities of the nodes. Cheaper nodes have higher capacity limitations and lower fault tolerance.

Scalability: The number of entities in the application may vary depending on the environment where it is implemented and on its task. Consider an application used for early detection of forest fire which is implemented in a huge forest such as the Amazons. Due to the fact that sensors should have a small transmission range on the order of a few meters in certain scenarios, the network must have on the order of thousands nodes in order to cover the whole area. In such a case, the algorithms running inside the network should scale well in parallel to the increasing number of nodes in a region maintaining the given task of operation properly. In other words, the network should adapt itself to changing node density without affecting the application performance. Scalability is a significant issue especially in outdoor applications.

Security: Ad hoc networking and wireless medium introduce many security flaws which make Cooperating Object networks open for various types of malicious attacks. The system may be threatened by unauthorised users trying to access the network. Also, there are security risks on the physical layer of the network. For example, jamming signals may corrupt the radio communication between the entities in the mission-critical networks.

Real-time/low end-to-end delay: End-to-end delay requirements are very stringent in real-time applications. For instance, in a manufacturing automation application actuation signal is required in real-time. Additionally, low end-to-end delay may be an essential requirement in some delay-sensitive applications such as target tracking.

Reliability: End-to-end reliability guarantees that the transmitted data is properly received by the receiving-end. In some applications end-to-end reliability may be a dominating performance metric whereas it may not be important for the others. Especially, in security and surveillance applications, end-to-end delivery is of highly importance.

Time Synchronisation: Associating the data coming from multiple objects to the same event, data aggregation, data fusion, target or event tracking tasks and cooperation make time synchronisation among communicating entities a key issue in many applications.

Below, common characteristics and requirements of each application domain are summarised.

Control and Automation (CA)

Networked embedded systems decrease the need for human power. The applications that fall into this category may be used in indoor or outdoor environments and they should provide the ability to enable distributed process control with ad hoc and robust networking in challenging environments. They include robotics, control and automation technologies, and artificial intelligence studies. The benefits of those applications may be explained by giving examples in the process manufacturing area where continuous research and implementation of new production technologies must be done in order to reduce waste, reduce time of operations and improve volumes. Through the use of networked sensors, robots, and process control algorithms, the performance of manufacturing process is readily increased.

In many control and automation applications, apart from basic sensor nodes, there are transportation systems and other entities which capable of making autonomous decisions. Also, those applications are mostly event-driven.

The applications in this category are certainly required to be working in real-time. Hence, having support for fault tolerance, end-to-end delay and synchronisation of the components are very important. Those applications are expected to decrease the need for interference of a human as much as possible, which is a fact that points to the importance of degree of automation. Also,the components in such a system may not be identical. Other issues are very dependent on the applications and their scenarios. So, some characteristics, like security, can be very important for some applications, while they are not an issue for some other applications.

Home and Office (HO)

Traditional life styles and habits have been changing with the emerging applications of embedded systems in home or office environments. Networked sensors with different tasks may arrange the room conditions such as temperature and light according to the needs of person. Furthermore, carbon monoxide sensors may be used to detect unsafe levels emitted from the heating systems. The security of an indoor environment may be increased as well by linking the home to private security companies.

In home and office applications, the components of the system behave depending on the state of the environmental context. Since almost all of the applications will be very personalised, security and privacy is highly important. Since the main inputs of the system will be originated from people, localisation is also important to state the location of the input to determine what to do. Such systems are inevitably designed to be context aware. The data to be processed or the feedback data of the system are expected to require a high bandwidth. Also, the system must be affordable by many people. A multi-hop network may not be necessary, also the topology and the routing will probably not be changed hence, mobility does not seem to be a challenging issue for this category.

Some state of the art projects in this category are: **Oxygen**, **ActiveBat**, **CORTEX's Smart Room**, **EasyLiving** and **Smart Surroundings**.

Oxygen [51] is an integrated vision and speech system uses cameras and microphone arrays to respond to a combination of pointing gestures and verbal commands developed by MIT Lab for Computer Science and Artificial Intelligence Lab. When you arrive at home, you say "I'm home", and the space comes alive. Lights flip on and music starts up on the stereo.

ActiveBat [2] is a context-aware ultrasonic indoor positioning system, developed by ATT Labs, Cambridge. ActiveBat transceivers are mounted on walls. They communicate with ActiveBat tags carried by objects in the environment and determine their positions. The principal of the location-finding system is the trilateration and speed-of-sound is used to estimate the distance between the objects and reference points from TOA measurements.

CORTEX's Smart Room[13] is an EU project. When someone enters into the sentient room, her identity is captured by some device in the room and it starts to behave intelligently in her preferred way.

EasyLiving [59] is a ubiquitous computing project developed by Microsoft Research. It aims at developing architecture and technologies for building intelligent environments.

Smart Surroundings [29], [34] is an EU project and the aim is to investigate, define, develop, and demonstrate the core architectures and frameworks for future ambient systems. It projects that people will be surrounded by embedded and flexibly networked systems that provide easily accessible yet unobtrusive support for an open-ended range of activities, to enrich daily life and to increase productivity at work. This idea requires ubiquitous computing. Although ActiveBat only provides location information to the users and do not promise to create an intelligent information, it is mentioned in this category in order to give some examples of the first home/office ubiquitous computing applications.

Logistics (L)

Logistics is a hot research field for new micro and nano technologies. Using Wireless Sensor Networks, a product can readily be followed from production step until it is delivered to the end user. For this purpose, it may cover a large geographical area and many diverse entities in order to establish communication between entities involved in the application. Therefore it may require high degree of distribution. Also, a manufacturer or company would like to minimise its expenses in order to increase its profit. For a corporation, especially the maintenance cost of a logistics application is a significant parameter in order to decide to establish it or not.

In logistic applications, it is not easy to let people be able to interfere with individual components of the system at any time for maintenance purposes. Therefore providing a fault tolerant system is important for this class. The system must also be able to serve to a scale of hundreds or thousands of inventories as well as it serves to tens of them. Mobility and localisation of the components are basic requirements of the applications in this category.

Due to the fact that many logistics applications benefit from the RFID technology, there are only a few scenarios mentioned in this category: **CoBIs-Collaborative Business Items**, **Zebra RFID Product Tracking**, **Smart Dust Inventory Control (SDIC)** and **Sun RFID Industry Solution Architecture**.

CoBIs-Collaborative Business Items [10] aims the usage of smart sensor technology in industrial/supply chain/sensitive settings. It will make it possible to implement networked embedded systems technologies in large-scale business processes and enterprise systems by developing the technologies for directly handling processes at the relevant point of action rather than in a centralised back-end system. Modeling embedded business services, developing the collaborative and technology frameworks for CoBIs with necessary management support, and investigating and evaluating CoBIs in real-world application trials in the oil and gas industry are the main objectives of the project.

Zebra RFID (Radio Frequency Identification) Product Tracking [75] is a method of remotely storing and retrieving data using devices called RFID tags. An RFID tag is a small object, such as an adhesive sticker, that can be attached to or incorporated into a product. RFID tags contain antennas to enable them to receive and respond to radio frequency queries from an RFID transceiver.

Smart Dust Inventory Control (SDIC) [63] project can be described with a simple scenario, based on a communication sequence. The carton talks to the box, the box talks to the palette, the palette talks to the truck, and the truck talks to the warehouse, and the truck and the warehouse talk to the Internet. Know where your products are and what shape they're in any time, anywhere. Sort of like FedEx tracking on steroids for all products in your production stream from raw materials to delivered goods.

In **Sun RFID Industry Solution Architecture** [67], the goal is to monitor the pharmaceutical supply chain in order to prevent drug counterfeiting and improve the health and safety of the people. It is based on the RFID technology.

In this category, several requirements such as localisation, scalability, end-to-end reliability are still challenging research issues.

Transportation (TA)

Applications of this class aim at providing people with more comfortable and safer transportation conditions. They offer valuable real-time data for a variety of governmental or commercial services. It is possible to design different scenarios of transportation applications. For instance, in one scenario cars may communicate with each other in order to organise traffic, whereas in another one traffic organisation may be done by installing static entities along highways and monitoring traffic conditions of roads. The desired result of these kinds of applications is to attain autonomous transportation systems.

The nature of transportation applications starts from being ad hoc, hence infrastructureless and mobile. One vital issue for those mobile components is localisation. The systems like car control or traffic applications are highly required to run in real-time. So, the synchronisation of the components and the end-to-end delay of the whole system are quite critical for such systems. Another issue that affects all these requirements is scalability. All the characteristic requirements we state here should be handled by taking care of different scales of the network.

There are many projects that aim at providing safer public transport. We have covered only following applications since the field of action does not necessarily involve a sensor network in the scenarios: **The Traffic Pulse**, **CORTEX's Car Control Project**, **Safe Traffic** and **CarTALK 2000**.

The Traffic Pulse is developed by Mobile Technologies, USA. This project is the foundation for all of Mobility Technologies applications. It collects data through a sensor network, processes and stores the data in a data centre and distributes that data through a wide range of applications. Digital Traffic Pulse Sensor Network has installed along major highways, the digital sensor network gathers lane- by- lane data on travel speeds, lane occupancy and vehicle counts. These basic elements make it possible to calculate average speeds and travel times. The data will then be transmitted to the data centre.

In **CORTEX's Car Control Project** [62], the implemented system will automatically select the optimal route according to desired time for reaching the destination, distance, current and predicted traffic, weather conditions, and any other information that will be necessary for the purpose. Cars cooperate with each other to move safely on the road, reduce traffic conditions and reach their destinations. Cars slow down automatically if there are some obstacles or they are approaching to other cars, speed up if there are no cars or obstacles. Cars automatically obey traffic lights. The principal target of this application scenario is to present the sentient object paradigm for real-time and ad hoc environments. It needs decentralised (distributed) algorithms.

Safe Traffic [68] project aims the implementation of an intelligent communication infrastructure. This communication system would provide all vehicles, persons and other objects located on or near a road with the necessary information needed to make traffic safer. In addition, all road-users should be provided with an accurate positioning device. The idea is simple: reducing the number of traffic-related deaths and injuries.

CarTALK 2000 [8] is a European Project focusing on new driver assistance systems which are based upon inter-vehicle communication. The main objectives are the development of cooperative driver assistance systems and the development of a self-organising ad-hoc radio network as a communication basis with the aim of preparing a future standard. To achieve a suitable communication system, algorithms for radio ad-hoc networks with extremely high dynamic network topologies are developed and prototypes will be tested in probe vehicles. These are mostly ad hoc applications with high mobility requirements.

Environmental Monitoring for Emergency Services (EM)

Environmental monitoring applications have crucial importance for scientific communities and society as a whole. Those applications may monitor indoor or outdoor environments. Supervised area may be thousands of square kilometres and the duration of the supervision may last years. Networked microsensors make it possible to obtain localised measurements and detailed information about natural spaces where it is not possible to do this through known methods. Not only communications but also cooperation such as statistical sampling, data aggregation are possible between nodes. An environmental monitoring application may be used in either a small or a wide area for the same purposes.

One of the first ideas of wireless sensor network concept is to design it to use the system to monitor environments, where humans cannot be present all the time. The main issue is to determine the location of the events. Such systems are to be infrastructure-less and very robust, because of the inevitable challenges in the nature, like living things or atmospheric events. Since the nodes are untethered and unattended in this class of applications, the system must be power efficient and fault tolerant. Long lifetime of the network must be preserved while the scale increases in order of tens or hundreds.

Environmental monitoring for emergency services is a typical domain which can benefit from networked tiny sensors. Several projects are underway: **GoodFood**, **Hogthrob**, **WaterNet**, **GlacsWeb**, **Sustainable Bridges** and **Smart Mesh Weather Forecasting (SMWF)**.

GoodFood [28] [32] project aims to develop the new generation of analytical methods based on Micro and Nano technology (MST and MNT) solution for the safety and quality assurance along the food chain in the agro-food industry.

Hogthrob aims the monitoring of sensor wearing sows. The use of sensor nodes on the animals could facilitate other monitoring activities: detecting the heat period (missing the day where a sow can become pregnant has a major impact on the pig production) and possibly detecting illness (such as a broken leg) or detecting the start of farrowing (turning on the heating system for newborns when farrowing starts).

The objective of **WaterNet** [70] was to provide the users, namely drinking water authorities, with a suitable technology and it has demonstrated the usability and appropriateness of newly developed methods in real-life applications.

GlacsWeb [26] project is being developed in order to be able to monitor glacier behaviour via different sensors and link them together into an intelligent web of resources. Probes are placed on and under glaciers and data collected from them by a base station on the surface. Measurements include temperature, pressure and subglacial movement.

The main goal of **Sustainable Bridges** [46] [27] is to develop a cost effective solution for detection of structural defects and to better predict the remaining lifetime of the bridges by providing the necessary infrastructure and algorithms.

Meteorology and Hydrology in Yosemite National Park is monitored by **Smart Mesh Weather Forecasting (SMWF)** system. Results are showing how, in some years, snowmelt may occur quite uniformly over the Sierra, while in others it varies with elevation. The number of commercial applications will increase when algorithms and paradigms satisfy challenging requirements such as scalability, robustness or power efficiency.

Healthcare (H)

Applications in this category include telemonitoring of human physiological data, tracking and monitoring of doctors and patients inside a hospital, drug administrator in hospitals etc. Merging wireless sensor technology into health and medicine applications will make life much easier for doctors, disabled people and patients. They will also make diagnosis and consultancy processes faster by patient monitoring entities consisting of sensors which provide the same information regardless of location and transition automatically from one network in a clinic to the other installed in patient's home. As a result, high quality healthcare services will get closer to the patients.

Health applications are critical, since vital events of humans will be monitored and automatically interfered. Heterogeneity is an issue, because the sensed materials will be various. Localisation is important because it is critical to determine where exactly the person is; if he carries a heart rate control device and it detects a sudden heart attack, there must be no mistake or no incapability for finding his location. However, since in most cases single-hop networks will be used and neither topology, nor the routing will be changed, mobility is not considered to be a challenging issue for this kind of applications. The delay between the source of the event, and the other end-point of the system is also important. The data has to be conserved as original, which points to reliability of transmission. Also the context is important. Supposing a sensing node must take into account that if a person is doing sports at that moment in order to tolerate higher heart rate differences. Although the idea of embedding wireless biomedical sensors inside human body is promising, many additional challenges exist; the system must be safe and reliable; require minimal maintenance; energy-harnessing from body heat. With more researches and progresses in this field, better quality of life can be achieved and medical cost can be reduced.

The projects, examined in this category are the following: **MyHeart**, **Ubiquitous Support for Medical Work in Hospitals (WS-MW)** and **AUBADE**.

MyHeart [48] aims at empowering the citizens to fight cardiovascular diseases by preventive lifestyle and early diagnosis. The first step is to obtain knowledge on a citizen's actual health status. In order to gain this information, continuous monitoring of vital signs is a must. The approach is therefore to integrate system solutions into functional clothes with integrated textile sensors. The combination of functional clothes and integrated electronics and process them on-body, we define as intelligent biomedical clothes. The processing consists of making diagnoses, detecting trends and react on it. MyHeart system is formed together with feedback devices, able to interact with the user as well as with professional services.

In the **Ubiquitous Support for Medical Work in Hospitals (US-MW)** [3] project, clinics can access relevant medical information and collaborate with colleagues and patient independent of time, place and whatever they are doing. Supposing a patient is lying on an interactive bed and on the other side there is a public wall-display in another room, clinic or hospital and a nurse will be having a real-time conference with a radiologist in this project.

In **AUBADE**, the goal is to create a wearable platform consisting of biosensors to ubiquitously monitor and recognise the emotional state of those kind of people in real-time.

Security and Surveillance (SS)

Sensors and embedded systems provide solutions for security and surveillance concerns. These kinds of applications may be established in varying environments such as deserts, forests, urban areas, etc. Communication and cooperation among networked devices increase the security of the concerned environment without human intervention. Natural disasters such as floods, earthquakes may be perceived earlier by installing networked embedded systems closer to places where these phenomena may occur. The system should respond to the changes of the environment as quick as possible.

Security and surveillance applications have the most number of challenging requirements. Almost all issues must be covered to develop such systems. These applications require real-time monitoring technologies with high security requirements. The environment to be observed will mostly be inaccessible by the humans all the time. Hence, robustness plays an important role. Additionally, maintenance may not be possible and power efficiency and fault tolerance must be satisfied.

Sample projects we covered in this category are: **Bio Watch**, **FloodNet**, **Vehicle Tracking and Autonomous Interception (VT-AI)**, **Monitoring Volcanic Eruptions with a Wireless Sensor Network (MVE-WSN)**, **PinPtr**, **COMETS**, **CROMAT** and **RISCOFF**.

Bio Watch [61] aims at providing early warning of a mass pathogen release, including anthrax, smallpox and plague. The primary goal of **FloodNet** [26], is to demonstrate a methodology whereby a set of sensors monitoring the river and functional floodplain environment at a particular location are connected by wireless links to other nodes to provide an "intelligent" sensor network.

Vehicle Tracking and Autonomous Interception (VT-AI) [60] is a networked system of distributed sensor nodes that detects an uncooperative agent called the evader and assists an autonomous robot called pursuer in capturing evader developed by University of California at Berkeley.

Monitoring Volcanic Eruptions with a Wireless Sensor Network (MVE-WSN) [72] is a wireless sensor network to monitor volcanic eruptions with low-frequency acoustic sensors developed by University of Harvard and University of North Carolina. The

PinPtr [43] system uses a wireless network of many low-cost sensors to determine both a shooter's location and the bullet's trajectory by measuring both the muzzle blast and the shock wave.

The **COMETS Project** [49, 11] (EU funded, IST-2001-34304) on real-time coordination and control of multiple heterogeneous UAVs includes the experimentation and demonstration of the system in forest fire detection and monitoring.

The **CROMAT Project** [15] on the cooperation of aerial and ground robots also considers applications in disaster scenarios. And finally, **RISCOFF** project studies and tests a sensor network system for forest fire detection and alarm signalling.

Tourism (T)

Everybody wants to feel safe and comfortable when he is in a new environment. When you visit a new country you want to find wherever you like to go without much effort. New micro and nano technologies may help tourists in a foreign environment. For example, sensors and hand-held devices may be a city guide, or may help people in an art museum. Location of the museums, restaurants and information about weather should be provided to the tourists.

Tourism oriented applications do not have high dependencies on the characteristics we mention here. However, they must be personalised. Those applications must be service and context aware and cost effective. They must also support mobility of the user. We present two sample studies: **Smart Sight** and **Cyber Guide**.

The **Smart Sight** [74] project is intended to translate from and to local language, handle queries posed and answer in spoken language, and be a navigational aid. The assistant is a "wearable computer" (consisting of a Xybernaut MAIV and a Thinkpad 600) with microphone, earphone, video camera, and GPS to determine the location of users. The system would have better knowledge of the environment than the tourist with accessing local database and the Internet.

The goal of **Cyber Guide** [16] is to provide information to a tourist based on her position and orientation. Initial prototypes of the Cyberguide were designed to assist visitors on a tour of the Graphics, Visualisation and Usability Centre during monthly open house sessions. The user will be able to see her current location and the demonstrators in her surroundings on a map.

Education and Training (ET)

Another emerging application domain of embedded systems lies in the area of education. It is possible to provide more attractive lab and classroom activities involving Cooperating Objects. Current activities aim at merging embedded systems into the education methods.

Although the requirements of education and training applications are highly dependent on the context, they all must be cost effective and affordable by many users. Since these applications are meant to ease the training of users by taking place of a living educator, they must have a high degree of automation. The other issues may be required more or less by the concept, context or type of the application.

Two projects are examined for education and training category: **Smart Kindergarten** and **Probeware**.

The **Smart Kindergarten** [65] project builds a sensor-based wireless network for early childhood education. It is envisioned that this interaction-based instruction method will soon take place of the traditional stimulus-responses based methods.

Probeware [64] is the other education and training project example which is an educational hardware and software tool used for real-time data acquisition, display, and analysis with a computer or calculator. It is also known as Microcomputer-Based Labs (MBL). When it is used with a calculator, it is known as Calculator Based Labs (CBL).

4.1.3 Current Trends

People working on research and development of wireless sensor networks, tend to use paradigms in the way which will lead them to gain maximum benefit. Lately, project groups are following a way to make the Cooperating Objects cooperate with each other in order to achieve a system as most autonomous as possible. In this manner, data centric system models play an important role. However, if a scenario needs more computing power, long system lifetime or high network traffic, service centric models or hybrid models are preferred.

For both in design stage and in deployment stage applications, in order to gain high network lifetime and getting rid of the necessity of renewing batteries in highly scaled networking applications, the topologies are planned to be multi-hop. This trend will eventually lead developers to build power-aware systems.

One of the most common trends is planning the projects to be as highest scalable as possible. This comes from a fact that almost all of the applications will be working on mediums where humans are not able to be present all the time and hence, covering the widest area possible. Also, localisation is another issue that researchers take into account, because of the fact that in a highly scaled network, the origin of any event is critical.

4.1.4 Conclusion

The survey of related projects have shown that, not counting academic test-beds, only few real wireless sensor applications are readily available today. Most applications are still at the prototyping or demo stage, mainly aiming at verifying node prototypes, identifying real requirements and understanding research challenges. Various enabling technologies such as energy efficient schemes for networking, time synchronisation, localisation and tracking are also being studied.

	CA	HO	L	TA	EM	H	SS	T	ET
Network Topology	AD	Single	AD	Multi	Multi	AD	Multi	AD	AD
Indoor/Outdoor	AD	In	Both	Out	Out	Both	Out	Both	In
Scalability	AD	X	√	√	√	X	√	AD	AD
Packaging	AD	X	AD	AD	√	X	√	AD	AD
Fault Tolerance	√	AD	√	AD	√	AD	√	AD	AD
Localisation	AD	√	√	√	√	√	√	AD	AD
Time Synch.	√	AD	AD	√	AD	AD	√	AD	AD
Security	AD	√	AD	AD	AD	AD	AD	X	AD
Infrastructure	AD	AD	AD	AdHoc	AD	AD	AdHoc	AD	AD
Prod-Maint.cost	√	√	√	X	√	X	AD	√	√
Mobility	AD	AD	√	√	AD	AD	AD	√	AD
Heterogeneity	√	AD	AD	√	AD	√	AD	X	AD
Autonomy	√	AD	X	AD	AD	AD	AD	AD	√
Power Awareness	AD	AD	√	AD	√	AD	√	AD	AD
Real-Time	√	AD	AD	√	AD	√	√	AD	AD
Context-awareness	AD	√	X	X	X	√	X	√	X
Reliability	X	X	√	X	X	√	X	X	X

Table 4.1: Characteristics of Cooperating Object Application Domains (AD: application scenario dependent)

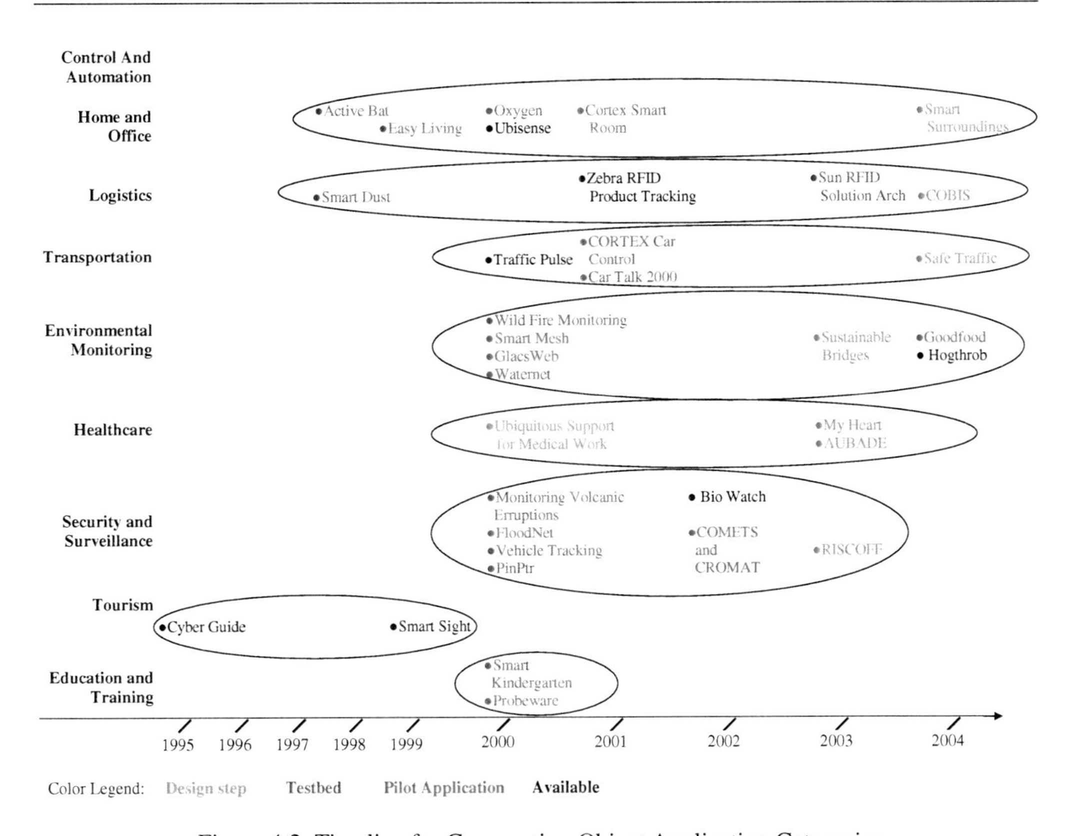

Figure 4.2: Timeline for Cooperating Object Application Categories

Currently deployed applications share some common characteristics: raw sensor data transmission over wireless connection, mostly data processing at collection points, simple routing schemes, best-effort data transport delivery. This shows that most application specific requirements are still challenging research issues. The field is similar to the situation of the Internet 30 years ago. The major difference is that the research is heavily oriented towards studying the field as application-specific. General-purpose solutions are yet to come. Another fact to point out is that the requirements and constraints of various applications are not yet fully understood, as a result, most of these applications are not ready for real world yet. Table 4.1 summarises the common and scenario dependent requirements and challenging issues of Cooperating Object application domains. Whereas $\surd$ stands for common characteristic of that category, X means that it is not relevant for that category. Application dependent issues are represented by 'AD'. As illustrated in Figure 4.2, almost all Cooperating Object projects have emerged in the last five years. The oval blocks show the time when an application category appeared to be interesting due to having a number of existing projects, but it says nothing about the amount of research performed in the field before and after this time.

At the time of this writing, we have been able to gather information about around 40 projects, of which 20% have to do with *Environmental Monitoring*. It seems that this area is the one that receives most attention nowadays. About 40% of the projects are in the *testbed* stage, whereas 23% are at the *design* stage, 17% at the *pilot application* stage and 20% at the *available* stage.

It is worth pointing out that this new paradigm has a potential of improving the quality of our lives considerably when the research in the field of common features is matured. With the progress on sensor fabrication techniques and multi-disciplinary research cooperation, we can expect that real-world cooperating object applications containing sensor network component will come to life in the near future.

4.2 Paradigms for Algorithms and Interactions

4.2.1 Introduction

The goal of this section is to provide an in-depth study and classification/taxonomy of the state of the art in terms of basic and advanced paradigms used for the design of algorithms and interaction patterns that apply to systems based on Cooperating Objects.

Since an exhaustive survey of the literature on Cooperating Object-based systems would be a daunting task and result in a very long document, we have decided to divide our search in four thematic areas that cover all the aspects of the application scenarios described in the previous section. These thematic areas are:

- *WSNEM:* Wireless Sensor Networks for Environmental Monitoring are characterised by a large number of stationary sensor nodes, disseminated in a wide area and one (or a few) sink nodes, designated to collect information from the sensors and act accordingly. WSNEMs can be deployed in greenhouses to monitor the environmental conditions, as the temperature and humidity of the air, and determine the actions to be taken, e.g., activating watering springs). Also, sensors can be used to detect any significant variations of the structure of building or bridges, in order to prevent crashes. Another relevant application field is that of Smart Environment, where active interaction among surrounding environment and people is made possible by the deployment of a distributed network of sensors. Depending on the user scenario, nodes can be either accurately placed in the area according to a pre-planned topology (e.g. for bridges monitoring), or randomly scattered over the area, except for a limited number of nodes that are placed in specific positions, for instance to guarantee connectivity or to act as beacons for the other nodes. Finally, a random topology is obtained when nodes are scattered in the area without any plan, as in the case of airplane dissemination of sensors over a forest or a contaminated area. Sensor nodes are often inaccessible, battery powered, prone to failure due to energy depletion or crashes. Furthermore, network topology can vary over the time due to the power on/off cycles that nodes go through to save energy. Generally, traffic in WSNEM flows mostly from sensors to one or more sinks and vice versa, and data may show strong spatial correlation. Since nodes are usually battery powered and not (easily) rechargeable, power consumption is a primary issue.

- *WSNMN:* A Wireless Sensor Network with Mobile Nodes is characterised by the use of mobile nodes in a Wireless Sensor Network, ranging from a network with only mobile nodes to a network with a trade-off between static and mobile nodes. The use of mobile nodes in sensor networks increases the capabilities of the network and allows dynamic adaptation with the changes of the environment. Some applications of mobile nodes could be: *collecting and storing sensor data in sensor networks* reducing the power consumption due to multi-hop data forwarding, *sensor calibration* using mobile nodes with different and eventually more accurate sensors, *reprogramming nodes* "on-the-air" for a particular application and *network repairing* when the static nodes are failing to sense and/or to communicate. In WSNMN, the mobile nodes can collect the data from the sensors and send it to a central station by using a long-range radio technology, thus acting as mobile sinks. Usually, the traffic between static and mobile nodes is low, whereas the traffic among mobile nodes and the central station is high. The mobile nodes can also process the information gathered from the static nodes in order to reduce the data traffic. Power Consumption is still an issue that has to be dealt with in WSNMB. However, mobile nodes can reduce the power consumed in multi-hop data forwarding. Furthermore, it is possible to have *energy stations* where mobile nodes can recharge their batteries.

- *ART:* Autonomous Robotic Teams (ARTs) make it possible to build more robust and reliable systems by combining unreliable but redundant components. The topology of ART is often pre-planned because the motion of the robots is always in some extent controlled and predictable. The data traffic among the robots is usually higher (images, telemetry, etc.) than the traffic among the nodes of a Wireless Sensor Network. Moreover, traffic patterns are more similar to the classic all-to-all paradigm considered in ad hoc networks. Furthermore, the cooperation and/or coordination of the robots requires very heavy data processing. Nevertheless, ARTs usually do not have severe energy constraints, since robots can autonomously recharge their batteries when a low level of energy is detected. Also, mixed solutions including solar panels are possible. Conversely to the other distributed systems here considered, ART can require real-time features, depending on the particular application considered.

- *IVN:* Intervehicle Networks or, in general, the cooperation among vehicles is a promising approach to address critical road safety and efficiency in IVN scenarios. For instance, coordinated collision avoidance systems could significantly reduce road accidents. The road safety becomes rather problematic in countries which heavily rely on the road network for transportation of goods.

  Inter Vehicular Communications (IVC) exhibit characteristics that are specific to scenarios of *high node mobility*. Indeed, despite the constraints on the movement of vehicles (i.e., they must stay on the lanes), the network will tend to experience very rapid changes in topology. Some of the IVN applications require communication with destinations that are groups of vehicles. Thus, the traffic pattern is predominantly multicast, in particular geocasting with physical areas of coverage, while the data rate is generally rather high (car-to-car voice/video connections, web surfing, etc.). Unlike typical Wireless Sensor Network deployment scenarios, vehicles can be instrumented with powerful sensors and radio to achieve long transmission ranges and high quality sensor data, since low-power consumption and small physical size are not an issue in this context.

Definition of Concepts

This section is aimed at providing a common understanding of the different concepts used throughout the text. For this purpose, we propose a short definition for each one of the concepts considered.

Paradigm In this context, the term *paradigm* refers to the methodologies and strategies that can be followed to address a problem and define its solution.

Algorithm An *algorithm* is the description of a step-by-step procedure for solving a problem or accomplishing some goals. Several different type of algorithms can be defined, according to the purposes they are designed for. In particular, this study deals with the following types of algorithms.

- *Medium Access Control:* MAC algorithms define the mechanisms used by the objects to share a common transmission medium.
- *Routing:* Generally speaking, a routing algorithm provides a mechanism to route the information units (usually data packets) from the source object(s) to the destination object(s).
- *Localisation:* Localisation algorithms are mechanisms that permit an object to determine its geographical position, either with respect to an absolute reference system or relatively to other objects in the area.

- *Data Processing:* Data processing includes both Data aggregation & Data Fusion techniques. Data aggregation algorithms are methods to combine data coming from different (and possibly heterogeneous) sources enroute, into an accounting record that can be then forwarded, reducing the number of transmissions, overhead and energy consumption of the system. A possible example is the aggregation of temperature and pressure data produced by two different sensors located in the same area into a single compounded packet that will hence delivered to an environmental-monitoring station.
  Data fusion algorithms are used to merge together information produced by different sources, in order to reduce redundancy or to provide a more synthetic description of the information. For example, the temperatures measured by several sensors located in a given area can be fused in a single average value for that area.
- *Synchronisation:* Synchronisation protocols allow nodes (or a subset of nodes that perform a common task) to synchronise their clocks, so that they all have the same time, or are aware of offsets of other nodes. Time synchronisation is essential for those numerous applications where events must be time-stamped. Moreover, protocol design is eased when nodes share a common clock (time division techniques for channel access, design of sleep/awake schedules, etc.). Time can be absolute (i.e., referred to an external, well-known measure of time), or nodes can agree a common time reference. This last case is useful in those case when time is needed to compare the occurrence of events.
- *Navigation:* Robots using sensor networks opens a new research area which includes the navigation of the autonomous robots using distributed information as a relevant issue. The navigation of the robot is possible even without carrying any sensor and just using the communications with the wireless sensor network. Furthermore, it should be mentioned that these algorithms can also be used for the guidance of people with a suitable interface.

Interactions The term *Interaction* refers to the exchange of information among objects that permits the realization of the coordination and cooperation of the objects:

- *Coordination:* is a process that arises within a system when given (either internal or external) resources are simultaneously required by several components of this system. In the case of autonomous robotic teams, there are two classic coordination issues to deal with: spatial and temporal coordination.
- *Cooperation:* can be defined as a joint collaborative behaviour that is directed toward some goal in which there is a common interest or reward. Furthermore, a definition for *cooperative behaviour* could be: given some task specified by a designer, a Cooperating Object system displays *cooperative behaviour* if, due to

some underlying mechanism (i.e., the "mechanism of cooperation"), there is an increase in the total utility of the system.

Taxonomy *taxonomy* consists in the classification of the concepts according to specific requirements and principles.

The taxonomy used in the following section is derived from the application requirements and characteristics defined by Section 4.1 and shown in table 4.2.

Topology	Scalability
Fault Tolerance	Localisation
Data Traffic Characteristics	Networking Infrastructure
Mobility	Node Heterogeneity
Power Consumption	Real-Time
Reliability	

Table 4.2: Applications Requirements and Characteristics Defined by Section 4.1

4.2.2 Classification of the Algorithms

In this section we will mainly refer to the algorithms covered by the Study on Paradigms for Algorithms & Interactions. For further details on the algorithms described, please refer to the specific sections of the study available from [12].

MAC algorithms

- *Topology:* MAC algorithms can be classified on the basis of the topology information they need to operate. Topology independent MAC algorithms, as those based on CSMA (MACA, MACAW, PAMAS) or SMAC and DBMAC, do not require any knowledge of the network topology. Other protocols, like SIFT and STEM, require nodes to have local information only, i.e., information regarding the nodes in their proximity. Finally, topology dependent protocols, such as TRAMA, assume nodes are aware of the entire network topology.
- *Scalability:* Generally, the performance of MAC protocols, in terms of medium access delay, is affected by the number of contending users. Contention-based access protocols, such as MACA, MACAW, PAMAS and so on, scale rather well with the number of nodes, when the traffic offered to the network is low. On the contrary, with high traffic loads, random protocols performance (in terms of medium access delay) worsens rather rapidly as the number of nodes increases. Contention-free MAC algorithms

(e.g., time-division based algorithms) scale better with high traffic loads, while for low traffic such solutions may incur in longer access delay than random algorithms.

- *Fault Tolerance:* In general, MAC algorithms are not affected by nodes failure, even though a certain performance loss may be experienced in case of topology dependent algorithms.

- *Localisation:* MAC algorithms can be classified in location-aware and location-independent. Location-aware solutions usually follow a cross-layer approach, since the location information is used both to manage the access to the medium and the forwarding of the information towards the intended destination (see GeRaF, Smart Broadcast). Some algorithms assume only that each node is acquainted (in some way) with its own spatial coordinates, others require the knowledge of the positions of the surrounding nodes only or of all the nodes in the network. Pure medium access algorithms are generally location independent. Location-aware algorithms are usually much more efficient that location-independent algorithms. However, they may turn out to be excessively sensitive to localisation errors. These aspects have not been sufficiently covered in the literature yet.

- *Data Traffic Characteristics:* Traffic characteristics may have a strong impact in MAC algorithms performance. Contention-based protocols usually show better performance in case of sporadic traffic bursts, while deterministic access mechanisms are more suited for handling periodic traffic generation patterns. At the state of the art, MAC algorithms do not consider the traffic flow patterns, i.e., the set of nodes that exchange data. An exception is represented by cross layer solutions that provide a integrated mechanism for both MAC and routing and are sometime designed according to the specific traffic flow pattern expected in the system.

- *Networking Infrastructure:* Generally, the presence of a network infrastructure permits to resort to contention-free MAC protocols based on polling strategies or resource reservation. However, most of the protocols covered by this study can be operated in absence of any networking infrastructure.

- *Mobility:* Mobility might represent an issue for MAC protocols for two reasons. First, mobility involves topology variations that may affect algorithms that need to tune some parameters according to the density of nodes in the contention area (SIFT, TRAMA, TSMA, MACAW). Second, MAC algorithms based on medium reservation mechanisms (MACA, MACAW) may fail in case of mobility, since the reservation procedures usually assume static nodes. For instance, algorithms based on the RTS/CTS handshake to reserve the medium may fail because either the corresponding nodes move outside the mutual coverage range after the handshake or external nodes get into the contention area and start transmitting without being aware of the medium reservation.

Nevertheless, many MAC algorithms considered in this study are capable of self-adapting to the topology variations in case of nodes mobility. Algorithms like TRAMA, TSMA and SMACS-EAR can still adapt to topology variations, but at the expense of the energy efficiency and the access delay.

- *Node Heterogeneity:* MAC algorithms for heterogeneous networks have not been yet investigated in the literature. Algorithms based on channel sensing (CSMA-based) provide some resilience to interference produce by other radio interfaces operating in the same frequency band and, hence, can be adopted in heterogeneous system. However, this solutions would not leverage on the nodes diversity. This topic is, indeed, still to be investigated in the literature.

- *Power Consumption:* Energy efficiency is considered in several MAC protocols, in particular in the case of wireless sensor networks. A typical method to reduce energy consumption is to let nodes alternate periods of activity and sleeping. Notice that such on/off cycles may be either managed independently of the MAC protocol or be part of it. For instance, CSMA, MACA, MACAW protocols do not explicitly consider the presence of such on/off cycles. Nevertheless, CSMA behaving is not affected by on/off cycles, while MACA and MACAW may fail since they assume nodes are always notified of the channel state. Protocols like PAMAS and SMAC, on the contrary, take into account the sleeping periods of the nodes, thus permitting a more efficient power management of the system. Usually, this is obtained at the cost of a higher complexity of the MAC protocol.

- *Real-Time:* Contention-based MAC protocols cannot usually provide any real-time guarantee. Conversely, contention-free algorithms, such as TSMA or TRAMA, are able to guarantee a given maximum access delay, which depends on the number of competing nodes.

- *Reliability:* Almost all the MAC algorithms considered in the study require explicit acknowledgement (ACK) of correct data reception from the receiver. Usually, in case of missing or negative ACK, the data link layer entity retransmits the data unit. However, the process is stopped when a given number of retransmissions is reached. In this case, the data unit is discarded. Hence, in general, MAC protocols can provide only limited reliability. Notice that, contention-based MAC algorithms are prone to transmission errors due to collisions, events that, on the contrary, never occur in contention-free algorithms. Therefore, contention-based algorithms are typically less reliable that contention-free ones.

The MAC algorithms taxonomy is summarised in table 4.3.

	CSMA	MACA MACAW PAMAS	SMACS-EAR	Sift	STEM	DB-MAC	TRAMA	TSMA	Energy-aware TDMA
Topology	Indep.	Indep.	Indep.	Only local	Only local	Indep.	Compl. top. knowledge	Only local	Compl. top. knowledge
Scalability	Partial (low traffic)	Partial (low traffic)	Partial (low traffic)	medium (low traffic)	medium (low traffic)	Partial (low traffic)	Good (high traffic)	Good (high traffic)	Good (high traffic)
Fault Tolerance	Resilient	Resilient	Resilient	Resilient	Resilient	Resilient	Partially Res.	Partially Res.	Partially Res.
Localis.	Not required	Not required	Not required	Not required	Not required	Not required	Not required	Not required	Not required
Data Traffic Characteristics	Better for sporadic traffic	Better for sporadic traffic	Better for sporadic traffic	Better for sporadic traffic	Better for sporadic traffic	Better for sporadic traffic	Better for periodic traffic	Better for periodic traffic	Better for periodic traffic
Netw. infrastruct.	None	None	None	None	None	None	None	None	None
Mobility	High resilience	Medium resilience	Medium resilience	Medium resilience	Low resilience	Medium resilience	Low resilience	Low resilience	Low resilience
Node Heterogeneity	None	None	None	None	None	None	None	None	None
Power Cons.	High	High	Medium (on/off)	High	Low (sleep)	Medium (Data aggr.)	Low (sleep)	Low (sleep)	Low (sleep)
Real-Time	Partial	Partial	Partial	Partial	No	Partial	Yes	Yes	Yes
Reliab.	Partial	Partial	Partial	Partial	Partial	Medium	High	High	High

Table 4.3: Taxonomy of the MAC Algorithms

Routing algorithms

- *Topology:* Routing algorithms can be differentiated on the base of the routing topology they realize. Usually, table-based algorithms create tree topologies, so that each node is the root of a routing tree towards each other node in the network.

  On-demand routing algorithms, on the contrary, realize point to point routing topologies, where a path from a node to its destination is created when needed (GedRaF, GAF, GEDIR). Cluster-based routing algorithms construct a hierarchical topology, in which some nodes are elected as cluster-heads (or coordinators) and forward data collected from their neighbours towards the final destination (LEACH, PEGASIS, TEEN). Request-driven routing algorithms aim at defining a path from possible multiple sources to the node that issues a specific data request. These algorithms lead to a star-shaped routing topology, where many paths originating from the source nodes converge to the destination node. Examples are Rumor Routing, Direct Diffusion routing.

- *Scalability:* Scalability is a important issue for routing protocols. Table-based protocols usually show scalability problems when the number of nodes (and, consequently, routing table entries) grows, in particular for systems with limited storing and computational capabilities (typically sensors networks). To alleviate this problem, many protocols resort to clustering techniques that, in turn, bring forth some control overhead (LEACH, TEEN, PEGASIS). Stateless algorithms have been introduced to cope with scarce storing capabilities, while maintaining good scaling properties. Typical examples are location-based algorithms, such as GeRaF, GAF, GEDIR, where nodes need to maintain the information regarding their own location and that of the destination node. Notwithstanding, the literature does not consider in detail the issue of distributing and maintaining the location information over the network.

  Algorithms that make use of broadcast packets to gather and/or diffuse topological information usually show scalability problems in large network due to the broadcast storm problem, unless broadcasting is obtained by means of specific broadcast-diffusion algorithm (Direct-diffusion, Rumor routing).

- *Fault Tolerance:* Usually, routing algorithms can adapt to topology variations due to nodes failure. However, the reaction to a topology variation due to nodes failure may require some time and, hence, bring some performance degradation. During this time, data can be delayed, duplicated or lost.

- *Localisation:* As seen for the MAC algorithms, also routing algorithms can be classified in location-aware and location-independent. Location-aware routing algorithms include the cross-layer solutions discussed in the classification of MAC algorithms, as GeRaF, and other pure routing algorithms, such as GAF, GEDIR, GEAR. Usually, location-aware routing algorithms assume that each node is acquainted (in some way) with its own spatial coordinates and those of the intended destination node. Hence, the next hop is determined in order to move the packet towards the destination.

  Data centric routing algorithms, such as Directed Diffusion, Rumor routing, SPIN, make use of broadcast techniques to disseminate and gather routing information and, therefore, do not require any localisation feature.

  Hierarchical routing algorithms, in general, are based on topological information, but do not require exact node localisation (LEACH, TEEN). Nevertheless, localisation may help the process of creating the cluster structure, thus resulting in better performance (VGA, TTDD).

- *Data Traffic Characteristics:* Data traffic characteristics may affect routing algorithms design. In Wireless Sensor Networks, for instance, spatial correlation among data generated by nodes in close proximity is exploited by cross layer solutions that merge routing and data processing functionalities (TEEN, VGA, COUGAR). Specific routing algorithms have been proposed for centralised traffic patterns, where information flows to and from a single central node (e.g., a sink node in Wireless Sensor Networks) and several peripheral nodes. Examples are SOP and MCFA.

- *Networking Infrastructure:* Most of the routing algorithms for cooperating objects are designed according to an ad-hoc paradigm. Therefore, solutions are completely distributed and do not require any backbone infrastructure.

- *Mobility:* Generally speaking, all the routing algorithms considered are able to cope with topology dynamic due to nodes mobility. However, most of them react to topology variations by dropping the broken paths and computing new ones from scratch, thus incurring in performance degradations. In particular, mobility may strongly affect cluster-based algorithms, due to the cost for maintaining the cluster-architecture over a set of mobile nodes. Routing algorithms specifically designed for networks with slow-mobile nodes are, for example, GAF and TTDD, which attempt to estimate the nodes trajectories. Other protocols that are well-suited for an environment where the sensors are mobile are the SPIN family of protocols because their forwarding decisions are based on local neighbourhood information.

- *Node Heterogeneity:* Node heterogeneity can be a winning feature to develop efficient routing algorithms, in particular for Wireless Sensor Networks with mobile nodes. Notwithstanding, the literature still lacks in solutions that leverage on nodes heterogeneity to enhance the routing process.

- *Power Consumption:* Power consumption is typically a very important issue in the design of routing protocols, since many cooperating-objects systems involve battery-powered units. Accordingly, several energy-efficient routing algorithms have been presented in the literature, in particular for Wireless Sensor Networks. The simplest manner to reduce power consumption is to allow each node to schedule sleeping periods. Therefore, routing protocols have to be designed to work also in the presence of on/off duty cycles (GAF, ASCENT).

  Other protocols to reduce the amount of information improve the energy efficiency of the system. Moreover, other techniques such as data aggregation, overhead reduction, cluster-heads rotation and so on can be used to reduce the energy wasting (LEACH, GEAR, PEGASIS, TEEN, HPAR).

- *Real-Time:* Routing algorithms that can provide tight constraint on the packet delivery time are rather seldom.

- *Reliability:* Most of the considered routing protocols cannot guarantee data reliability, especially when the network is rarely populated. Some routing algorithms, such as GeRaF, GAF, GEDIR, SPIN, may fail to discover a path in case of connectivity holes within a connected network. Other algorithms can, instead, guarantee delivery if source and destination nodes are connected (GOAFR, SPAN, LEACH, PEGASIS). Broadcasting-based algorithms, such as Rumor routing and Direct diffusion, generally offer high reliability thanks to the capillary diffusion of the routing control packets.

The routing-algorithm taxonomy is summarised in table 4.4.

	GeRaF	GAF SPAN	SPIN	LEACH PEGASIS	TEEN	HPAR	Rumor Direct-Diffusion
Topology	Point-to-point	Point-to-point	Star	Hierar.	Hierar.	Hierar.	Star
Scalability	Good	Good	Good	High	High	High	Low
Fault Tolerance	High	High	High	Medium	Medium	Medium	Medium
Localisation	Required	Required	Not required	May help	May help	May help	Not required
Data Traffic Characteristics	Irrelevant	Irrelevant	Relevant	Relevant	Relevant	Relevant	Relevant
Networking infr.	None	None	None	None	None	None	None
Mobility	High resilience	High resilience	High resilience	Medium resilience	Medium resilience	Medium resilience	High resilience
Node Heterogeneity	None	None	None	None	None	None	None
Power Cons.	Low (on-off)	Low (sleep)	Medium (data aggr.)	Low (clust.)	Low (clust.)	Low (clust.)	High
Real-Time	No	No	No	No	No	No	No
Reliability	Partial	Partial	Medium	Medium	Medium	Medium	Medium

Table 4.4: Taxonomy of the Routing Algorithms

Localisation algorithms

- *Topology:* Localisation algorithms are used to infer the geographical position of a node by elaborating the signals received from position-aware nodes (beacons/landmarks). The precision of the estimation is usually strictly dependent upon the placement of the beacons. Therefore, the network topology may have effects on the performance of most localisation algorithms. For example, in case of range-free approaches, inhomogeneous nodes density may lead to incorrect distance estimate (Centroid, DV-Hop).

- *Scalability:* Localisation algorithms are usually scalable with the network population. However, if the geographical extension of the network increases, a higher number of beacons may have to be deployed (APIT). Even if a multilateration approach is adopted, relaxing the need for direct beacons visibility, an increasing of the average number of hops from the beacons leads to localisation errors accumulation (DV-Host, DV-Dist). The complexity of the localisation algorithms, as well as the preciseness of the estimation, usually increase with the number of beacons (AHLoS). To conclude, the localisation algorithms, in general, might show scalability problem with the number of nodes that populate the network.

- *Fault Tolerance:* The localisation algorithms are usually tolerant to the dead of some nodes, given that they are not beacons. Failure of beacons is, instead, particularly critical for localisation algorithms performance (Centroid, DV-Hop, DV-Dist). Also, malfunctioning nodes, for instance nodes with defecting HW, may have an impact on localisation errors and on localisation error propagation (DV-Hop, N-Hop TERRAIN, AHLoS).

- *Localisation:* Localisation algorithms require, in general, a suitable disposition of the beacon nodes in the network area. Estimation may also be refined by using the positioning information estimated by the surrounding nodes.

- *Data Traffic Characteristics:* To estimate the position, sensor nodes use the control packets sent by beacons that contribute to the network load (DV-Hop, DVB Distance, N-Hop Terrain). In the case of networks with mobile nodes, furthermore, the position estimation might be improved by increasing the beacons frequency. Hence, a tradeoff between localisation accuracy and network load can arise. Furthermore, some localisation algorithms might make use of data packets sent by position-aware nodes to adjust their position estimation. In this case, regular or periodic data traffic exchange involving position-aware nodes can improve the performance of the localisation mechanisms without introducing extra control traffic.

- *Networking Infrastructure:* In general, localisation algorithms make use of infrastructures. In the specific, satellite-based positioning mechanisms obviously require a complex satellite network infrastructure. More generally, localisation algorithms require the presence of a network infrastructure that hosts beacon nodes, whose positioning information is disseminated over the network. Localisation algorithms that aim at providing only relative positions of a node in a network, on the contrary, do not require any settled infrastructure.

- *Mobility:* In general, nodes mobility increases the localisation error. However, in some context, mobile nodes capable of accurate position estimation might be used to disseminate positioning information over a network of elementary static nodes.

- *Node Heterogeneity:* The use of the satellite-based positioning systems is not always possible, for it increases the cost of the nodes and the power consumption. The heterogeneity of nodes play a fundamental role in these scenarios, since a bunch of localisation-enabled nodes might be exploited by the other nodes of a network to derive an estimation of their position (APIT).

- *Power Consumption:* Localisation schemes increase the power consumption. In particular, the use of satellite-based schemes is very expensive in terms of power consumption in some contexts (such as Wireless Sensor Networks). This cost might be reduced by installing a limited number of such devices in the network (beacons) and by using localisation algorithms to estimate the position of the other nodes in the network. Clearly, in this case the energy consumption is due to the control packets exchange (DV-Hop). Other localisation strategies encompass the use of ultrasonic transceivers (Cricket, AHLoS) that, however, determine further energy consumption.

- *Real-Time:* In general, satellite-based positioning system are capable to provide quasi real-time localisation service. On the contrary, localisation algorithms that are based on the elaboration of beacon signals are not suitable for strictly real-time applications, since, in general, they require the reception of several control packets to reduce the estimation error. On the other hand the problem of whether a real-time application can be supported or not becomes an issue only for mobile networks of cooperating objects. In many application scenarios in which nodes are instead static the localisation process can be performed at the network set-up, reducing its costs and allowing to use the different types of algorithms independently of the real-time constraints of the application.

- *Reliability:* Reliability of the localisation algorithms depends of the number and position of the beacon nodes, possibly the number of hops over which the localisation error propagates, the presence of malfunctioning or malicious nodes. Malicious nodes are nodes whose purpose is to compromise the correct operation of the network. Such nodes can provide for example wrong ranging estimates or wrong information on their own position to other nodes, affecting the other nodes localisation accuracy or the ranging estimate accuracy (DV-Hop, N-Hop TERRAIN, AHLoS). Ways to detect and filter the information provided by malfunctioning or malicious nodes have to be provided.

The localisation-algorithm taxonomy is summarised in table 4.5.

	Range-free				Range-Based		
	Centroid	DV-Hop	APIT	Monte Carlo	AHLoS, N-Hop Multi-lat.	DV-DIST, HOP-TER.	AFL
Topology	Symmetric	Uniform	Generic	Generic	Generic	Generic	Generic
Scalability	Good	Limited	Good	Good	Good	Limited	Very Good
Fault Tolerance	Partial	Partial	Good	Good	Partial	Partial	Good
Localisation	Coarse	Good / Coarse	Good / Coarse	Good	Good	Coarse	Good
Data Traffic Characteristics	Local	Flooding	Local	Local	Local	Flooding	Local
Networking infr.	Required	Required	Required	Required	Required	Required	Not required
Mobility	Fragile	Fragile	Required	Robust	Fragile	Fragile	Partially robust
Node Heterogeneity	Landmarks	Landmarks	Landmarks	None	None	Landmarks	None
Power Cons.	Low	High	Medium	Medium	High	High	Low
Real-Time	No	No	No	Partial	No	No	No
Reliability	Partial	Partial	Partial	Partial	Partial	Partial	Medium

Table 4.5: Taxonomy of the Localisation Algorithms

Data Processing

- *Topology:* The network topology might play an important role on the design of specific data processing. The perfect knowledge of the network topology, for instance, can be used to determine the position of the better collector nodes. Moreover, if the topology is pre-planned, nodes with more computational capabilities can be displaced in strategic position. On the contrary, in case of random topology placed, the choice of more suitable aggregation points have to be taken in a distributed manner and can be less efficient. Regarding the organisation of the network structure, the data processing techniques can lie on different communication topologies. Most known algorithms (Directed Diffusion, LEACH, PEGASIS, TAG and TiNA) run over tree-based or hierarchical structures. Differently, other schemes such as Synopsis Diffusion and Tributaries and Deltas organise the network in a concentric ring structure.

- *Scalability:* Scalability is an important goal in the design of efficient data processing techniques especially in large and dynamic networks. Existing data processing techniques based on the construction of some aggregation tree are less scalable than the multipath schemes due to the high cost to maintain the organisation of the network.

This characteristic is accentuated in large or dynamic networks where adding or removing some nodes from the tree structure heavily impact in the performance of the algorithms. On the contrary, multipath solutions offer a good scalability especially due to the local and distributed functionalities.

- *Fault Tolerance:* Data processing, in general, is performed in order to reduce the intrinsic data redundancy that might characterise some cooperating-objects scenarios (e.g.,WSNEM). On the other hand, data redundancy may assure a higher reliability in case of sensor failure, connectivity holes and so on. Hence, a tradeoff between fault tolerance and redundancy reduction has to be cut. More in detail, in case of low packet loss probability, tree-based algorithms achieve better performance because they are able to minimising the number of transmissions to deliver data reducing as much as possible the redundancy. On the contrary, the multipath schemes preserve some data redundancy so that perform better in case of high packet loss probability. There are also some hybrid approach such as Tributaries and Deltas which are able to tune their behaviour according to the link conditions.
- *Localisation:* In some case data aggregation techniques require information about the location of nodes. Nevertheless almost all data processing techniques do not require any type of localisation methods.
- *Data Traffic Characteristics:* The design of data processing techniques is strongly correlated to the specific considered application. In some cases, for instance in WSNEM, may be useful to perform data aggregation as near as possible to the data sources due to the high redundancy among data collected in the same spatial region. In other cases, data processing can be performed along the path, for instance to merge information flows directed to a common destination.
- *Networking Infrastructure:* Data processing can take advantages from the presence of networking infrastructure. For instance, access points can play the role of data collectors and perform any type of simple data processing before forwarding information to the end destination. But at this time, none of the proposed algorithms make use of existing network infrastructure.
- *Mobility:* Mobility might improve the efficiency of data aggregation techniques. For instance, nodes with controlled or predictable motion can be driven all over the network to collect, process and store data generated by static nodes. On the contrary, mobility can affect the performance of the data processing schemes based on the aggregation tree.
- *Node Heterogeneity:* Data processing may exploit objects with higher storage and computational capabilities as aggregation centres, in order to convey and process data from less-powerful objects displaced in a common region. Moreover, it is necessary

to take into account the different node capabilities when the aggregation functions or the data structures are designed. For instance the proposed Q-digest structure can be used to store data with a different degree of precision according to the storage capabilities of the nodes.

- *Power Consumption:* Data processing techniques are, in general, implemented to limit the energy consumption by reducing the amount of transmitted data or the network overhead. Nevertheless, the required processing power contributes to deplete the energy resources of collector nodes. This aspect, however, is rarely considered in the literature and needs further investigation. At this time, aggregation functions implemented by algorithms such as Directed Diffusion, LEACH, TiNA, TAG are very simple (in general they are statistical function) and they do not require additional power consumption.

- *Real-Time:* Data processing techniques usually involve time delay for gathering and processing many data units from different sources. Consequently, such techniques might not be guarantee real-time requirements. This drawback is independent by the algorithm because it derives form the need to collect more than one packet before aggregating data and sending a new packet.

- *Reliability:* Data processing techniques are usually reliable, though, as mentioned, they might incur in large delays that could affect the utility of the delivered data for the final node. Also in this cases, multipath strategies could guarantee a higher reliability than the tree-based schemes.

The data processing algorithm taxonomy is summarised in table 4.6.

Navigation algorithms

- *Topology:* Information about the topology can be used to improve the accuracy of the localisation algorithms, providing a better performance of the navigation algorithm. The localisation has a more significant impact on the performance of the path computation and following algorithm (PACFA) when comparing with others. Therefore, information about topology can improve the accuracy in the navigation in general, and especially with PACFA.

- *Scalability:* The navigation algorithms presented in the study use local information provided by the Wireless Sensor Network and these algorithms have only been tested with one mobile node. Then, the scalability would depend on the capability of the communication protocol to support it. Furthermore, the use of a team of mobile nodes would involve other considerations such as the coordination among them for optimal covering of an area or collision avoidance for example. Those aspects could have a significant impact in the amount of information exchanged.

	Direct Diffusion	LEACH, PEGA-SIS	TAG, Cougar, TiNA	Synopsis Diffusion	Tributaries and Deltas
Topology	Tree-based	Tree-based	Tree-based	Ring	Hybrid
Scalability	Low	Low	Low	High	Medium
Fault Tolerance	Low	Low	Low	High	High
Localisation	None	None	None	None	None
Data Traffic Characteristics	Relevant	Relevant	Relevant	Relevant	Relevant
Networking infr.	None	None	None	None	None
Mobility	High resilience	High resilience	High resilience	Low resilience	Medium resilience
Node Heterogeneity	None	None	None	None	None
Power Cons.	Low	Low	Low	High	Medium
Real-Time	No	No	No	No	No
Reliability	Medium	Medium	Medium	High	High

Table 4.6: Taxonomy of the Data Processing Algorithms

- *Fault Tolerance:* The information provided by Wireless Sensor Networks used in the navigation algorithms improve the fault tolerance w.r.t. mobile nodes that only use the information from sensors installed on board.
- *Localisation:* Localisation of both the static nodes of the Wireless Sensor Network and the mobile nodes is required for most of the navigation algorithms. The performance of the potential field guiding (POFA) and probabilistic navigation (PRONA) algorithms is more robust to localisation errors than PACFA. If the localisation of the nodes is not provided "a priori", the navigation algorithm should also involve a position estimation.
- *Data Traffic Characteristics:* The navigation algorithms involve the exchange of a large amount of data among the static and the mobile nodes. These data should be updated at a rate which depends on the speed of the mobile node. PRONA does not require a "explicit" computation of the path and therefore a lower amount of data is involved. Finally, the increasing in the information exchanged due to the navigation algorithm can not exceed the capacity of the sensor network.
- *Networking Infrastructure:* The navigation algorithms found in the literature only use local information from the nodes close to the mobile node. Then, there is not any special requirement regarding the networking infrastructure.

- *Mobility:* This is an intrinsic characteristic of these algorithms, that can be applied to guide a robot or a person with a suitable interface in a given environment.
- *Node Heterogeneity:* It is also an intrinsic characteristic of these algorithms due to the fact that both static and mobile nodes are present. Even among the mobile nodes different characteristics, such as the locomotion system, are possible.
- *Power Consumption:* Power consumption of the nodes is increased due to the higher information exchange rate required during navigation. PACFA and POFA involve a first stage to compute the path, so a higher power consumption is required.
- *Real-Time:* Real-time requirements mainly depend on the speed of the mobile node. On the other hand, the navigation algorithms found in the literature are designed considering negligible delays in the information exchange among the nodes.
- *Reliability:* Besides the general reliability issues in Wireless Sensor Networks, the reliability of the mobile platform itself must be also considered.

The navigation algorithm taxonomy is summarised in Table 4.7.

	POFA	PACFA	PRONA
Topology	Non-relevant	Relevant	Non-relevant
Scalability	Not considered	Not considered	Not considered
Fault Tolerance	High	High	High
Localisation	Less required	More required	Less required
Data Traffic Characteristics	Very high rate	Very high rate	High rate
Networking Infrastructure	None	None	None
Mobility	Intrinsic	Intrinsic	Intrinsic
Node Heterogeneity	Intrinsic	Intrinsic	Intrinsic
Power Consumption	Very high	Very high	High
Real-Time	Relevant	Relevant	Relevant
Reliability	High	High	High

Table 4.7: Taxonomy of the Navigation Algorithms

4.2.3 Timeline

Figure 4.3 presents the different concepts that have been classified in this section according to the year of their appearance in the literature. The figure offers a one-look view of the evolution of current trends.

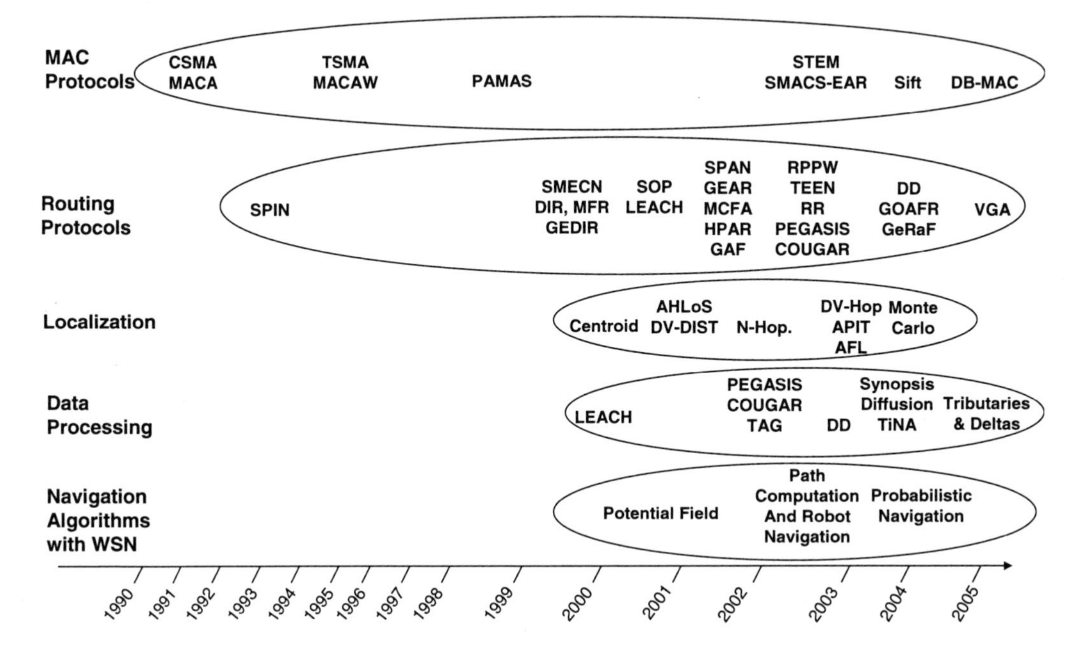

Figure 4.3: Timeline of the Literature

4.2.4 Conclusion

This section and its associated longer document available from [12] was conceived to provide an in-depth analysis of the literature regarding cooperating-object systems and, hence, to identify a set of algorithms and architectures that could form a common framework for the next generation of Cooperating Object-based systems. Unfortunately, the Cooperating Object umbrella encompasses systems with static and energy-limited nodes, very low duty cycles and very flexible delay constraints, as well as systems with autonomous mobile nodes, no energy supply problems, and strict requirements in terms of communication delay and reliability, as clearly arises from the analysis of the reference thematic areas considered in the study.

A first conclusion that can be drawn from this study, however, is that Cooperating Objects may present irreconcilable discrepancies, which make hardly feasible the definition of a unified approach for the design of algorithms and solutions for this type of systems. Nonetheless, it is possible to identify cross-layer approaches, dynamic hierarchical architecture, location-based solutions, asynchronous communication paradigms, wireless communication and heterogeneity as common *trends* that are transversal to the plethora of different design approaches.

4.3 Vertical System Functions

4.3.1 Introduction

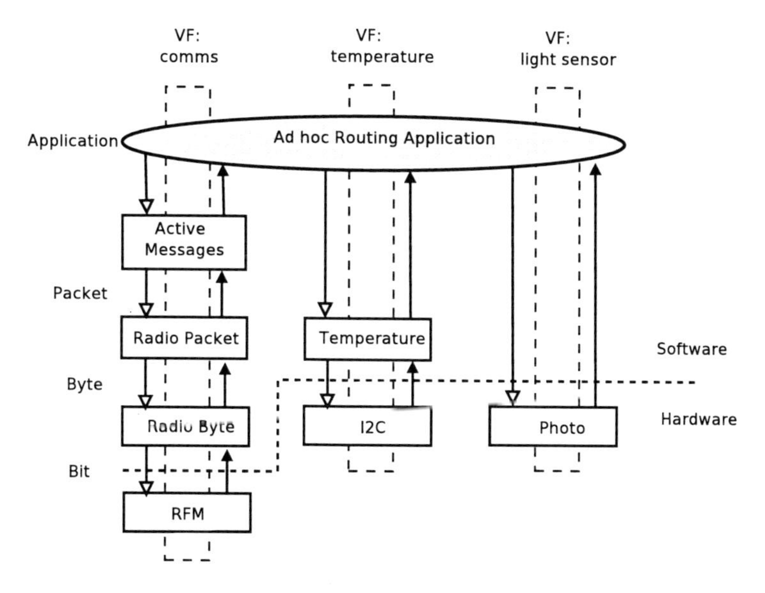

Figure 4.4: Example of Vertical Functions: Temperature and Light Sensing Application

The current operating systems proposed for Wireless Sensor Networks and Cooperating Objects cannot offer all the required functionality to the applications. Thus, vertical system functions (VFs) are defined in this summary as the functionality that addresses the needs of applications in specific domains and in some cases a vertical function also offers minimal essential functionality that is missing from available real-time operating systems.

Figure 4.4 introduces a simplified architectural view of an application-example originally discussed in [31]. The goal is to monitor the temperature and light conditions of an area and periodically transmit their measurements to a central base station. In this example, there are three VFs that applications may 'invoke': the communication subsystem (core system function), the light and temperature sensors (application specific functions). Each VF is represented in the diagram by a vertical stack of components. We envisage that *standardised APIs* will create mechanisms for linking applications to vertical functions.

The architectural framework introduced in Figure 4.4 refers to a stand-alone cooperating object. We believe that in practice there will be collections of Cooperating Objects in constant interaction to accomplish a preassigned goal. In some application scenarios, VFs will be implemented through a chain of software components that may or may not be within the operational boundaries of a single Cooperating Object but rather distributed in the network. For example, a VF that is responsible for collecting temperature readings of rooms in an office building needs distributed coordination among Cooperating Objects located in each room in order to implement the intended functionality.

4.3.2 Types of Vertical System Functions

This section presents a summary of the vertical system functions surveyed in the longer study available from [12]. The Applications and Application Scenarios section (section 4.1) discussed a list of characteristics and requirements of selected cooperating objects applications. Table 4.8 relates each of these requirements to the VFs.

We will use the order below to discuss the relevant vertical system functions in the sections that follow:

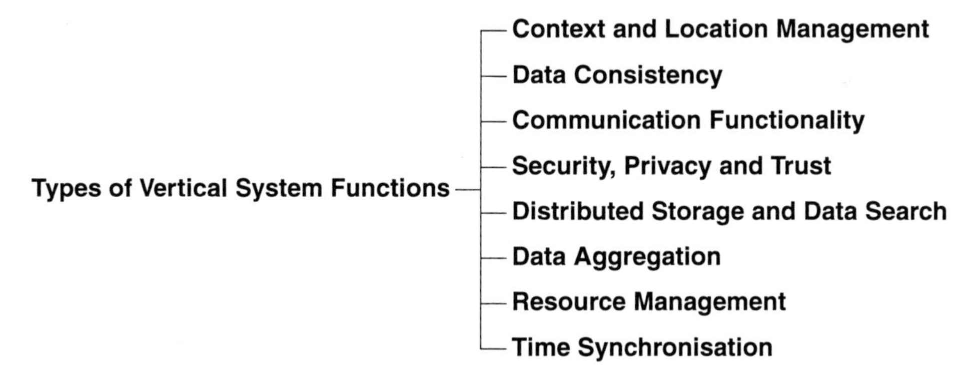

4.3.3 Context and Location Management

Distributed Cooperating Objects systems are designed to measure properties of the physical world. They are, therefore, suitable for gathering the context of an entity, which is the

Characteristic/requirement	Vertical System Function
Network Topology – topologies to support either single or multihop communication.	Communication; Distributed storage and search
Scalability – necessary system support to growing number of Cooperating Objects.	Distributed storage and search; Data consistency
Fault Tolerance – provides the mechanisms for supporting the resilience of the system in failures.	Data consistency; Communication; Security
Localisation – determining a node's location is the basic required functionality for various applications.	Context and location management
Time Synchronisation – Cooperating Objects need to establish a common sense of time.	Time synchronisation
Security – this is pervasive and must be integrated into every system component to achieve a secure system.	Security, privacy and trust
Data Traffic Characteristics – the system should provide support for various types of application traffic.	Communication; Distributed storage and data Search; Data aggregation
Networking Infrastructure – Cooperating Object networks can be infrastructured or infrastructure-less (ad hoc).	Context and location management; Communication
Mobility – the physical components of the system in some applications may be static whereas in others, the architecture may contain mobile nodes.	Context and location management; Communication
Node Heterogeneity – most of Cooperating Object applications include nodes that have distinct hardware and software factors.	Data lookup mechanisms; Data consistency, Data aggregation
Power Awareness – power consumption is one of the performance metrics and limiting factors almost in every Cooperating Object application.	Communication; Resource management
Real-time – the system delay requirements are very stringent in real-time applications. The broad meaning of delay in this context comprises the system data processing and network delay.	Resource management
Reliability – guarantees that the data is properly received by the applications.	Data consistency; Communication; Security, privacy and trust

Table 4.8: Application Requirements and Vertical System Functions

information that can be used to characterise its situation. Individuals, locations, or any relevant objects can be such entities. Since a reasonable amount of data is collected in large systems, context management systems are needed to handle them. Such systems can separate applications from the process of sensor processing and context fusion. To achieve precise actuation and detailed analysis of collected measurement data, however, the location of sensors and actuators need to be known.

Changes in context may trigger actions to influence the monitored entity. Specialised actuators, for instance, may be programmed to control pipe valves when a fluid pressure reaches a certain threshold. Classical context management systems use infrastructure-based directories to store the information, for example Aura and Nexus.

Current systems focus on the gathering of sensor data using quality of service specifications, the management of context as cross-layer data in the network stack or more general for the whole node, the storage of information in the network of cooperating objects at calculated geographic locations (Geographic Hash Table).

Adaptation of applications to the context is crucial. This can either be done application-driven, i.e. the application decides which actions should be taken, or system-driven, i.e. the system manages the adaptation transparently. The latter class has drawn attention in the last few years.

A few adaptive middlewares or frameworks have been proposed including Impala which is based on finite state machine or TinyCubus which tries to select the best set of algorithms based on several parameters, policies, and different adaptation strategies.

Location services for mobile ad hoc networks only offer limited context information – in this case the position of mobile objects. A scalable, distributed location service is required.

Prior to storing the location in a such a service, the location has to be determined. Small cooperating objects usually do not have a GPS device, so different methods have been developed in the last few years. Two basic approaches are commonly used: having distances to three objects of which the location is known, the own location can be calculated. The other possibility is to measure the angle to two known objects. Since most cooperating objects are equipped with omni-directional antennas, a very accurate measurement of the angles is not feasible.

4.3.4 Data Consistency and Adaptivity in Wireless Sensor Networks

The benefits of having several nodes in a Wireless Sensor Network mostly come from the fact that many nodes simultaneously monitor the same physical area. Nodes can be put into sleep mode without any loss of precision in the network. This results in energy conservation increasing, therefore, the network lifetime.

The reliability of the system is also improved with several sensor nodes. This scenario, however, raises issues with data inconsistency which may occur due to various reasons - for instance inherent imprecision associated with sensors, inconsistent readings and unreliable data transfer, just to name a few.

This vertical function provides the functionality to ensure consistency of the sensor data at various system abstraction levels:

- Data consistency may mean that data retrieved from a location in the sensor network should be consistent with data sent to the same location.
- Data consistency may also mean that all sensors sensing the same physical phenomenon should more or less agree on the measured value.
- In a rule-based system, data consistency may mean that all actuators agree on the action that needs to be taken.

For state monitoring, describing the creations between states and the events that trigger state-change through a set of rules and predicates over events and their parameters has proved to be popular.

Also hybrid distributed algorithms similar to the one proposed by Sahni et al. [58], which is executed by every sensor using the measurement ranges received from the remaining sensors monitoring the same region as data, accompanied with the sensor's own measurement have received considerable attention.

Introducing, defining and implementing the concept of *collaboration* at various levels of abstraction have recently been used to address the issue of data consistency.

4.3.5 Application-level Communication

The communication vertical function offers the functionality for any pair (or group) of devices to exchange information. Different types of communications can be performed: one-to-all, one-to-many, many-to-one, many-to-many. If we consider the case of wireless sensor networks with a single sink, one to all or one to many communications are needed for query dissemination, while many to one communication is explored to gather data at the sink.

Communication in Wireless Sensor Networks has been mainly data centric and attribute-based. This means that more than addressing a specific cooperating object, the communication infrastructure should be able to deliver data to and from groups of Cooperating Objects which share a set of attributes specifying the destination/source address of the information.

Data-centric and attribute-based data dissemination may lead to the selection of different communications overlays depending on the values to be reported. A given overlay interconnects cooperating objects of the same 'group' sharing a common set of attributes (attribute-based routing). Attribute-based routing could leverage the burden of delivering queries to objects which cannot answer the query. Also attribute-aware communication infrastructures may optimise sensor data fusion by selecting routes which maximise the chances of aggregating a given type of data, overall decreasing the network load and energy-consumption.

Important VF parameters which should be included in a query are the time constraints and accuracy with which a given query needs to be answered. This can be regarded as new concept of *quality of service* requirements that remains to be explored.

4.3.6 Security, Privacy and Trust

Cooperating Objects and Wireless Sensor Networks are usually placed in locations that are accessible to everyone – also to attackers.

In a Cooperating Object, security is pervasive and it must be integrated into every component to achieve a secure system. Components designed without security can become a point of attack. However, specific vertical functions to enforce security are available for applications.

Since fully tamper resistant devices are hard to build and would cost too much money, protocols for sensor networks have to be designed in a way that they tolerate malfunctioning/attacking nodes while the whole sensor network remains functional.

Routing protocols were developed that are resilient to black-hole attacks or that use efficient symmetric key primitives to prevent compromised nodes from tampering with uncompromised routes consisting of uncompromised nodes. It also deals with a large number of DoS attacks.

SIA, a framework for secure information aggregation in large sensor networks, uses random sampling mechanisms and interactive proofs to verify that the answer of an aggregator is a good approximation of the true value.

Encryption is the basic technique for securing and authenticating transmitted data. Using asymmetric cryptography on highly resource constrained devices is often not possible due to delay, energy and memory constraints. With symmetric cryptographic methods, two communicating cooperating objects need a common key. Secure Pebblenets use a shared key for the whole network but since the compromise of one single node leads to a security failure of the whole network pairwise approaches have been developed in the last years. SPINS uses a central base station to establish new pairwise session keys.

Using cooperating objects, especially sensor networks, humans can be observed which leads to privacy problems. Encryption tackles overhearing and data coarsening ensures that no conclusions can be drawn from the data to a single person. The usage of sensor networks to spy on individuals cannot be met with technology alone, but only with a mix of societal norms, new laws, and technological responses [52].

With several cooperating objects contributing to a common goal, it is necessary to assess the reliability of the information provided by an individual cooperating object. With respect to services and transactions, trust has been researched for several years. For data-centric and fully distributed architectures research has just started. A distributed voting system is proposed in [9] where votes can be cast against misbehaving nodes until all other nodes refuse to communicate with this node.

4.3.7 Distributed Storage and Data Search

Efficient storage and querying of data are both critical and challenging issues. Especially in Wireless Sensor Networks large amounts of data are collected by a high number of tiny sensor nodes. Scalability, power and fault tolerance constraints make distributed storage, search and aggregation of these sensed data essential.

It is possible to perceive a Wireless Sensor Network as a distributed database and run queries which can be given in SQL format. These queries can also imply some rules about how to aggregate the sensed data while being conveyed from sensor nodes to the query owner.

Data querying systems in general have two major components; interest/data dissemination and query processing and resolution. Query resolution usually involves data aggregation for energy efficient processing.

Interest and Data Dissemination Protocols for data dissemination are designed to efficiently transmit and receive queries and sensed data in Wireless Sensor Networks. In this subsection, we briefly explain five of these protocols.

- *Classic Flooding:* a node that has data to disseminate broadcasts the data to all of its neighbours.
- *Gossiping:* this technique uses randomisation to conserve energy as an alternative to the classic flooding approach. Instead of forwarding data to all its neighbours, a gossiping node only forwards data to one randomly selected neighbour.
- *SPIN:* this protocol is based on the advertisement of data available in sensor nodes. When a node has data to send, it broadcasts an advertisement (ADV) packet. The nodes interested in this data reply back with a request (REQ) packet. Then the node disseminates the data to the interested nodes by using data (DATA) packets. When a node receives data, it also broadcasts an ADV, and relay DATA packets to the nodes that send REQ packets. Hence the data is delivered to every node that may have an interest.
- *Directed Diffusion:* in SPIN the routing process is stimulated by sensor nodes. Another approach, namely Directed Diffusion, is sink oriented. A sink is the name given to the central node responsible for gathering data from all the other nodes in Directed Diffusion where the sink floods a task to stimulate data dissemination throughout the sensor network. While the task is being flooded, sensor nodes record the nodes which

send the task to them as their gradient, and hence the alternative paths from sensor nodes to the sink are established. When there is data to send to the sink, this is forwarded to the gradients. One of the paths established is reinforced by the sink. After that point, the packets are not forwarded to all of the gradients but to the gradient in the reinforced path.

- *LEACH:* is a clustering based protocol that employs randomised rotation of local cluster heads to evenly distribute the load among the sensors in the network. In LEACH, the nodes organise themselves into local clusters, with one node acting as a local cluster head. LEACH includes randomised rotations of the high-energy cluster-head position such that it rotates among the various sensors in order not to drain the battery of a single sensor. In addition LEACH performs local data fusion to compress the amount of data being sent from the clusters to the base station.

Query Processing and Resolution When a query arrives at a sensor node, it is first processed by the node. If the node can resolve the query, the result of the query is disseminated. This approach is one of the simplest ways of resolving and processing a query. Sensor nodes usually take advantage of collaborative processing to resolve queries so that smaller number of messages are transmitted in the network. Queries can be flooding-based where a query is flooded to every node in the network. Alternatively they can be expanded ring search based on the assumption that a node does not relay a query that it can resolve. Currently available query processing systems are summarised below:

- *TinyDB:* is a query processing system for extracting information from a network of TinyOS sensors. TinyDB provides a simple SQL-like interface to specify the data along with additional parameters such as the rate at which data should be refreshed much as in traditional databases. Given a query specifying data interests, TinyDB collects data from nodes in the environment, filters and aggregates them. TinyDB does this via power-efficient in-network processing algorithms. Some key features of TinyDB areas follows: TinyDB provides metadata management, provides a declarative query language, supports multiple query resolution on the same set of nodes and supports different levels of in-network aggregation. It also includes a facility for simple triggers, or queries that execute some command when a result is produced.

- *COUGAR:* is a query layer for sensor networks which accepts queries in a declarative language that are then optimised to generate efficient query execution plans with in-network processing which can significantly reduce resource requirements.

- *Active Query Forwarding scheme (ACQUIRE):* aims at reducing the number of nodes involved in queries. In this scheme each node that forwards a query tries to resolve it. If the node resolves the query, it does not forward it further but sends the result back. Nodes collaborate with their n hop neighbours, where n is referred to as the look ahead parameter. If a node cannot resolve a query after collaborating with n hop neighbours, it forwards it to another neighbour. When n equals to 1, ACQUIRE carries out flooding in the worst case.

4.3.8 Resource Management

It enables high-level system primitives to hide unnecessary low-level details. It enables the application to be independent of the actual underlying distributed system. It should also address the dynamic nature of available resources, such as variable network bandwidth. Because the system is built of error-prone components, failures should be handled as normal and not as exceptions.

Research have focused on the schemes to replace the traditional strict modularisation or layering design with cross-layered design, which is energy-efficient. However, most of the important cross-layer adaptation frameworks that have been proposed with the assumption of mobile devices that are richer in resources than wireless sensor nodes. Adaptation frameworks designed specifically for Wireless Sensor Networks are emerging.

Operating systems for wireless sensor networks with data centric architecture and especially designed for limited memory are considered to form one of the main trends in the area of adaptation in Wireless Sensor Networks. Due to increasing interest in having adaptive behaviour on a Wireless Sensor Network-wide scale, recently attentions have been paid to extending the concept of data centric architecture to the whole network.

4.3.9 Time Synchronisation

Applications need to establish a common sense of time among the cooperating objects participating in their sensing and actuation goals. Such a functionality can be offered through a time synchronisation vertical function.

Forest fire monitoring is a scenario that requires not only the information of whether there has been an indication of fire but also where and when this is happening. The collected sensor data provides the basis for the decision making process which may trigger actions to be taken in the monitored environment in order to address abnormal circumstances. Thus, the true time of the observed events is crucial to the prompt action of fire fighting.

An important question to ask is whether the Network Time Protocol (NTP) currently in use on the Internet could be adopted without further modification in low cost wireless sensor networks. This protocol relies on an external time reference to synchronise the top layer of time servers called Stratum 1 servers.

As the NTP protocol was originally developed considering other design issues, some of the challenges that arise in Cooperating Objects and Wireless Sensor Network scenarios cannot be addressed if the protocol is adopted unmodified in sensor network applications.

One of the issues is energy efficiency as the external source of time (e.g GPS) tends to consume more power than other components of a sensor node and does not work in indoor areas which have no line-of-sight to the satellites.

Research has focused on the design issues of self-configuring protocols assuming that NTP cannot be directly used in cooperating objects and Wireless Sensor Network applications. The latest research results address the time synchronisation problem with approaches that do not rely on GPS time signals. As researchers realise that low-power and low-cost GPS devices are becoming commercially available, the research trend may shift to hybrid protocol designs where GPS is used as the primary source of time. Such information is then disseminate throughout the network to non-GPS nodes. How such time servers ought to be organised in the network (e.g. hierarchical such as NTP) would be an important research question.

The characteristics of the network topology in ad-hoc wireless sensor networks may introduce severe delays because of disconnected parts of the network. To compensate for such uncertainties, the time synchronisation protocol needs to be *tolerant* to the message delay irrespective of its dominant source (e.g. processing, transmission).

4.3.10 Timeline

Figure 4.5 summarises the timeframes of the approaches taken to address the issues in each vertical function. They mainly relate to the research efforts in the field of wireless sensor networks. As it can be observed most of the research are fairly recent, ranging between 2000 and 2006.

4.3.11 Conclusion

Vertical Functions represent an area of research that cannot be considered in isolation, but together with the requirements of the application and application scenarios, the system requirements and the properties of the algorithms themselves.

In this section, we have considered vertical functions such as context and location management, data consistency and adaptivity, communication functionality, security, privacy and trust, distributed storage and data search, resource management and time synchronisation.

As it can be extracted from the current trends and the timeline depicted in Figure 4.5, the study of vertical functions and active research in this field is relatively new when compared to other algorithms and paradigms in the field of Cooperating Objects. It is due to the optimisation requirements of the applications themselves that these types of interactions (different from traditional layered architectures) is needed and starts playing an important role that will continue as long as the field remains as heterogeneous and varied as it is today.

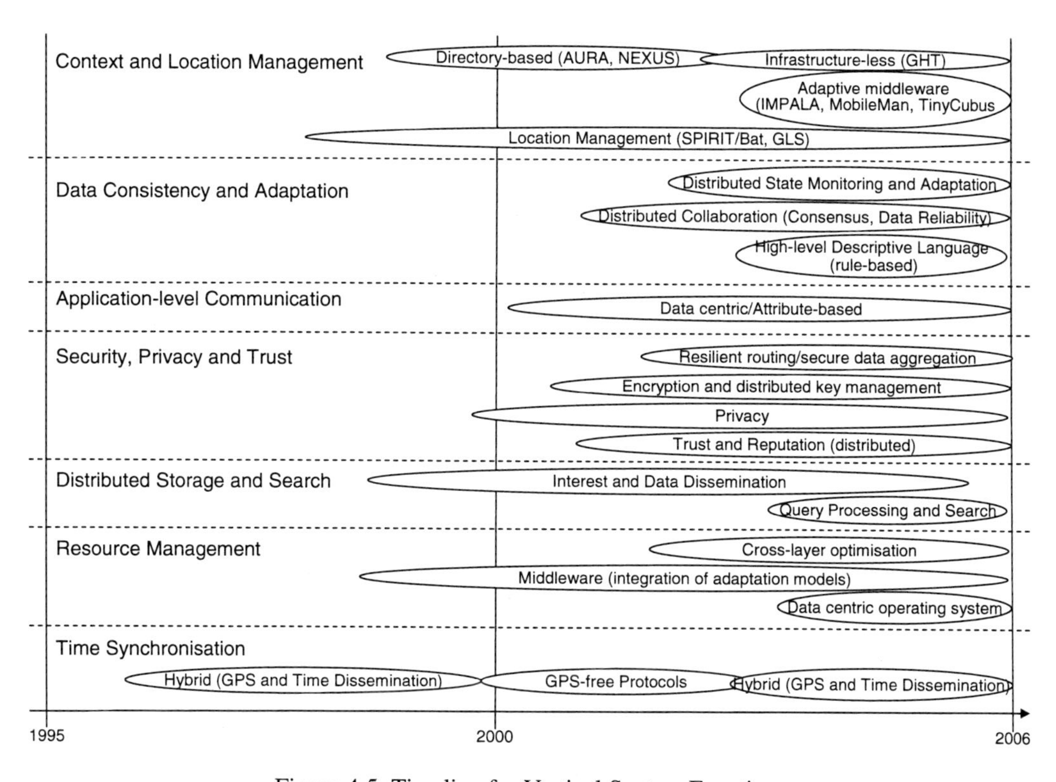

Figure 4.5: Timeline for Vertical System Functions

4.4 System Architecture and Programming Models

4.4.1 Introduction

Key to the successful and widespread deployment of cooperating objects and sensor network technologies is the provision of appropriate programming abstractions and the establishment of efficient system architectures able to deal with the complexity of such systems. Programming abstractions shield the programmer from the "nasty system details" and allow the developer to think in terms of the concrete application problem rather than in terms of the system. This is also true for traditional distributed systems, where numerous software frameworks and middleware architectures are crucial to perform an integrated computing task. Such frameworks and middleware are based on programming models such as distributed objects or events. These conventional and successful distributed programming abstractions can, however, not be simply applied to cooperating objects or sensor networks, due to the differences existing among the latter and the former systems.

We will refer to a programming model as "a set of abstractions and paradigms designed to support the use of computing, communication and sensing resources in an application" and to a system architecture as "the structure and organisation of a computing system, as a set of functional modules and their interactions".

In this section, we present the current trends that emerge in programming models and system architectures for cooperating objects and network sensors. Canonical examples of the paradigms identified in these trends are also presented. Then, we give some of the missing topics in the current researches and state of the art, which seems important to us as ground for further investigations. For more details, refer to the complete study on system architectures and programming models available from [12].

4.4.2 Trends and Canonical Examples

Traditional systems and abstractions

This trend tries to re-use existing system architectures and programming abstractions and apply them to the context of sensor networks and cooperating objects. An obvious advantage is to leverage on existing programmer knowledge, which can be re-used directly. In this trend, we find operating systems derived from desktop computing platform, and standard communication facilities. They are briefly described in the following sections.

Operating Systems Traditional desktop operating systems have been scaled to run on embedded systems, but they still require too much memory (e.g., about 4-5 MBytes for Windows XP Embedded) to run on smaller cooperating objects. Their advantage is that the development of applications is much easier since the OS is based on known concepts. For hand-held devices, special OS with a significantly smaller footprint have been designed, e.g. Palm OS, Symbian OS, or Windows CE. These OS support only basic functionality such as task scheduling, interrupt handling, and power management.

Traditional embedded operating systems are designed for real-time systems, e.g. eCos, QNX, or XMK (eXtreme Minimal Kernel), whereas the currently dominating operating systems targeted for wireless sensor networks are not real-time systems, e.g. TinyOS [31], Contiki [19], or Mantis [1]. Two recent operating systems for sensor nodes that attempt to provide real-time support are Timber [35] and DCOS [18]. Possibly, the need for real-time systems also in sensor networks becomes more evident when more real applications are deployed.

TinyOS and the kOS (kind-of or kilobit) operating system do not provide any multithreading, Contiki provides multithreading as a library for those applications that explicitly require multithreading, and Mantis is a layered multithreaded operating system. Without multithreading, programming is done with an event-based paradigm which fragments the control flow of programs and, thus, makes them harder to understand. Contiki's protothreads allow to write programs without multithreading even with block calls. A trend towards multithreading and procedural programs is recognisable.

TinyOS is compiled with the application into one single binary and is, therefore, not any more modular in binary form. No code can be loaded dynamically. In contrast, a Contiki system is partitioned into the core and the loaded programs. Thus, Contiki has support for loading individual programs from the network. DCOS and Mantis support dynamic reprogramming over the radio as well. As Contiki, SOS emphasises the notion of dynamically loadable modules. With real sensor networks deployed in large numbers and/or in inaccessible places, dynamic code update is crucial. More systems are expected to be developed in this area.

Standard Communication Facilities In the context of cooperating objects, many applications do not use new abstractions, and rely on standard communication facilities to support cooperation between objects. Typically, they use some form or resource or service discovery system in order to acquire and update the computing context, and explicit communication between objects once the context has been determined.

Examples of resource discovery systems include JINI [69], or Bluetooth SDP.

Depending on the context, objects cooperating may involve various abstraction levels of communications:

- Packet or message level (MAC, Bluetooth RFCOMM, UDP, ...)
- File or object level (OBEX, embedded web server, ...)
- Service level (RPC, RMI, web services, ...)

These approaches consist in re-using client-server and distributed objects paradigms: each cooperating object hosts one (or more) server(s) enabling other objects to use its services, either using traditional client/server interactions or through remote method invocation. The Cooltown system is typical of this approach, where cooperating objects (printers, etc.) are running embedded web servers, accessed by other objects (PDAs etc.) running web clients.

Other well know examples are JINI and CORBA's ORB.

Virtualisation

In this trend, the approaches attempt at providing easier programming environment through homogeneity and virtualisation of physical nodes. Typically, they take the form of virtual machines and dedicated language or scripts (such as Java, Tcl, or SQL). We list below some of the well-known approaches that fall in this category.

The concept of virtual machines is well-known since the early sixties and used to indicate a piece of computer software able to shield applications from the details of an underlying hardware or software platform. A virtual machine offers applications a suite of "virtual" instructions and attends to map them to the "real" instruction set actually provided by the underlying real machine. In this way, the virtual machine abstraction can mask differences in the hardware and software layers below the virtual machine facilitating code and data mobility.

A manifold of virtual machines have been developed to fulfill many different requirements and purposes, but only a few have been investigated and designed explicitly for systems of cooperating objects.

In the context of sensor network, we can cite *Maté*, developed at the Intel Research Laboratory, MagnetOS [4, 40], SensorWare [7]. In the context of mobile embedded systems, we can cite Scylla [66], and finally Java [39] (PJava and MIDP) which is very widespread on personal mobile devices.

Database View A system composed by a manifold of entities like simple sensor nodes, complex devices or just common everyday objects, each one of them endowed with sensing capabilities, may be regarded as a distributed database, where users can issue *SQL*-like queries to have the system performing a certain task or delivering required data.
In this perspective the system appears just as a collection of sensors, whose readings need to be adequately combined and queried, since manually retrieving and processing data from thousands of devices is substantially impracticable. Unfortunately, traditional data-processing techniques and operators from the database community cannot be applied one by one to systems like sensor networks, since the traditional assumptions about reliability, availability and requirements of data sources cannot be extended to simple sensors.

In the last few years, many researchers have noted the benefits of a query-like interface to sensor networks and some systems were developed following the "database philosophy", like *COUGAR* ([5]), *IrisNet* or *TinyDB* ([42]). The most successful among the proposed systems is *TinyDB*, a query processing system for extracting information from a network of tiny wireless sensors developed at the Intel Research Laboratory Berkeley in conjunction with the UC Berkeley.

In the context of Cooperating Objects, the database view is used in the PerSEND system to support proximate collaborations between PDAs. In this model, a federated view of a database is maintained from the data available on each node. The database model is relational, and the system proposes an SQL-like interface to the applications. The database view is dynamic in the sense that it directly reflects a physical context made of the set of near-by objects. As objects moves, the context evolves and the data associated to the objects are added or deleted from the database view. This system relies on a decentralised architecture, using only peer to peer communications (one-hop) over short distance wireless interfaces.

Data Distribution and Addressing

Sensor networks are systems which map data over a particular geographical area, hence approaches focusing data distribution and addressing are important.

Database View and Data Service Middleware A first approach is the database view (see 4.4.2, with systems like TinyDB and Cougar). Unlike these systems, data Service Middleware (DSWare) [37, 38] does not need a sink node that translates queries. It is a specialised layer in the network stack that performs the integration of various real-time data services for sensor networks. It supports reliable data storage, caching of popular data in the regions that need it, group management, and group-based decision making. In DSWare, a confidence value is associated with each decision that is made based on a prespecified confidence function.

Mobile Code and Mobile Agents Another approach is *Mobile Code* and mobile agent, which consist in a software program transmitted from an entity to another through a network to be executed at the destination. An examples is SensorWare [7] which uses Tcl scripts. *Remote Evaluation*, *Code-on-demand* and *Mobile Agents* are the three basic paradigms that are encompassed in the notion of *Mobile Code*. *Mobile Agents* in particular, represent mobile code that autonomously migrates between entities, and are therefore well suited for the implementation of distributed applications. An example from the ubiquitous computing domain is Shop Navi, one of the first systems to introduce Mobile Agents to support pervasive computing interactions.

The notion of *Mobile Agents* may be easily seen as an efficient programming strategy for sensor networks, since sensing tasks may be specified as mobile code scripts that may spread across the sensor network carrying collected data items with them. These scripts may be injected into the network at any point and are able to travel in the network and distribute itself where and when necessary.

Addressing based on Group or Spatial Properties Finally, we can identify approaches where the addressing is based on group or spatial properties.

The concept of *clustering* is a well-known paradigm in the field of distributed systems and ad-hoc networks, and offers a suitable programming abstraction for systems collecting a manifold of complex devices, which cooperate and coordinate to reach a common goal [20].

In a sensor networks, for example, nodes that share some neighbourhood relationship can organise themselves in *groups* within which nodes can efficiently communicate and share local resources and that constitute single addressable entities for the programmer. Examples of this concept are *Hood* [73] and *Abstract Regions* [71].

Accessing network resources using *spatial references*, in the same way as in traditional imperative programming variables are accessed using memory references, is the underlying idea of "Spatial programming", a space-aware programming paradigm particularly suitable for distributed embedded systems. In this view, a networked embedded system is seen as a single virtual address space and applications can access network resources by defining a *spatial reference*, i.e. a pair {*space:tag*}, where *space* indicates the expected physical location and *tag* a property of the demanded network resource.

A system that supports the "Spatial programming" abstraction has been implemented by Borcea et al. using Smart Messages, a software architecture that recalls many concepts and constructs typical of mobile agents [6].

In a *Spatial Programming* view, physical cooperating objects are used as data symbols, that cover a given geometrical shape, and the physical space is used as a way to structure information and processing such as in SPREAD [14]. The idea of the approach is to annotate existing interactions of physical entities with computing actions. These actions are triggered according to geometrical conditions, in particular physical proximity.

Self Organisation and Regulation

Many sensor networks applications benefit when application needs can be expressed in terms of global goals, and the network is able to self-configure accordingly to the application requirements and goals.

Virtual Markets The market-based approach offers a very expressive and intuitive way to model and analyse typical distributed control problems, as well as guidelines for the design and implementation of distributed systems. This methodology, however, has been also proposed as a generic programming paradigm for distributed systems and addressed as *market-oriented programming*. This approach regards modules in the distributed system as autonomous agents holding particular knowledge, preferences and abilities and the distributed computation may be implemented as a market price system [47]. This abstraction is particularly suited to model systems where a manifold of different devices need to cooperate and coordinate in order to reach a common goal in a globally efficient way. Under this point of view, the system is seen as a virtual market where agents (i.e., single devices) act as self-interested entities, which regulate their behaviour to achieve the maximal profit with the minimal costs (resource usage) considering globally-known price information (set in order to achieve a globally efficient behaviour).

In the context of the cooperation of personal communication devices, some studies are investigating the use of economic like regulation systems to enforce a global objective. For example, in the IST Secure project, each object manages trust values in other objects by recording the positive or negative outcomes of each interaction with other objects.

Role-based Abstractions and Adaptive System Software In role based approaches, sensor nodes take on specific functions or roles in the network without manual intervention. These roles may be based on varying sensor node properties (e.g., available sensors, location, network neighbours) and may be used to support applications requiring heterogeneous node functionality (e.g., clustering, data aggregation).

The task of assigning appropriate roles to sensor nodes is an implicit part of many networking protocols. The concept of role assignment is also a common function of middleware platforms for sensor networks.

MiLan [30] allows sensor network applications to specify their quality needs for sensor data and subsequently makes adjustments on specific properties of the sensor network to meet the sensor network needs while considering available resources. Impala [41] goes beyond that and supports on-the-fly application adaptation based on parameters and device failures which allows to improve the performance, reliability and energy-efficiency of the system. The adaptation capability is static since it is based on a finite state machine where different protocols are assigned to different states and conditions on the parameters represent the transitions.

A more general approach was presented with the TinyCubus project [44, 45] whose goal is the creation of a generic reconfigurable framework for sensor networks. The Data Management Framework of the TinyCubus provides a set of data management and system components. Each component is classified according to its suitability to several parameters. The framework is then responsible for the selection of the appropriate implementation based on current parameters contained in the system. TinyCubus also includes a Tiny Cross-Layer Framework which provides a generic interface to support the parameterisation of components that use cross-layer interactions and a Tiny Configuration Engine which distributes and installs new code in the network. The generic aspect of the role assignment problem is developed by Frank et al.[22] and [57].

Global Information Space

Several approaches tend to support cooperation of objects by providing a common global information space, which is used for communication and/or synchronisation. We already mentioned discovery systems (see 4.4.2), which provide a way to centralise the description of the resources and services available to an object.

Other approaches that can be identified in this trend are shared information space, such as tuples space proposed in LIME [53]. In these approaches, a shared information space is maintained by the objects. The shared information space may be centralised on one object, or distributed over a set of objects.

Support for Control Synchronisation

As any form of distributed processing, cooperating objects need support synchronisation and as one may expect, most approaches use a synchronisation paradigm. We identified two main ways: event-based approaches, and synchronisation through shared information space.

Event Detection *Events* are a natural way to represent state changes in the real world and in distributed systems, giving rise to model applications as producers, consumers, filters, and aggregators of events. Regardless of the specific scenario, "interesting events" may represent both node-internal occurrences (timeouts, message sending or receiving) than specific sensing results. Thus, the application can specify interest in certain state changes of the real world ("basic events") and upon detecting such an event, a sensing-device sends a so-called *event notification* towards interested applications. The application can also specify certain patterns of events ("compound events"), such that the application is only notified if occurred events match this pattern [56], [36].
The Event Detection paradigm is particularly well suited to provide a programming abstraction for sensor networks applications.

Shared Information Space Tuple space systems like LINDA provide synchronisation using a producer/consumer paradigm defined for tuple reading and writing operations. Control flows in applications programmed with this model are easy to understand, as it is directly driven by the data flow of the application.

4.4.3 Timeline

Figures 4.6 and 4.7 present the different concepts that have been classified in this section according to the year of their appearance in the literature. The figure offers a one-look view of the evolution of current trends.

4.4.4 Identified Gaps

This section describes the missing topics that were identified at this time, and that can be considered as important topics for further researches.

Real-time Aspects

Real-time aspects will become more important in sensor networks in the future. Thereto, real-time operating systems for sensor networks have to be developed.

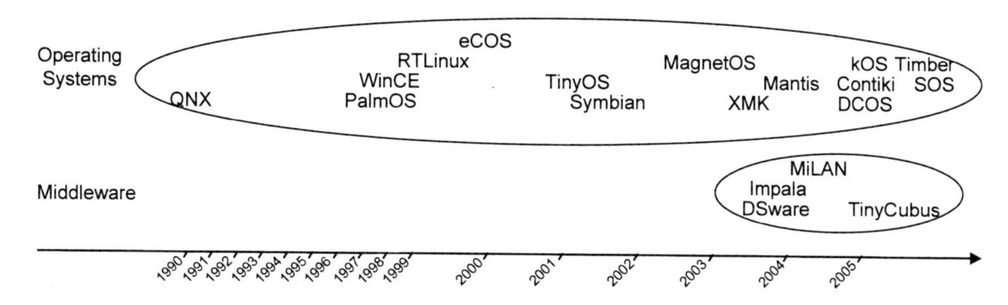

Figure 4.6: Timeline of System Architectures

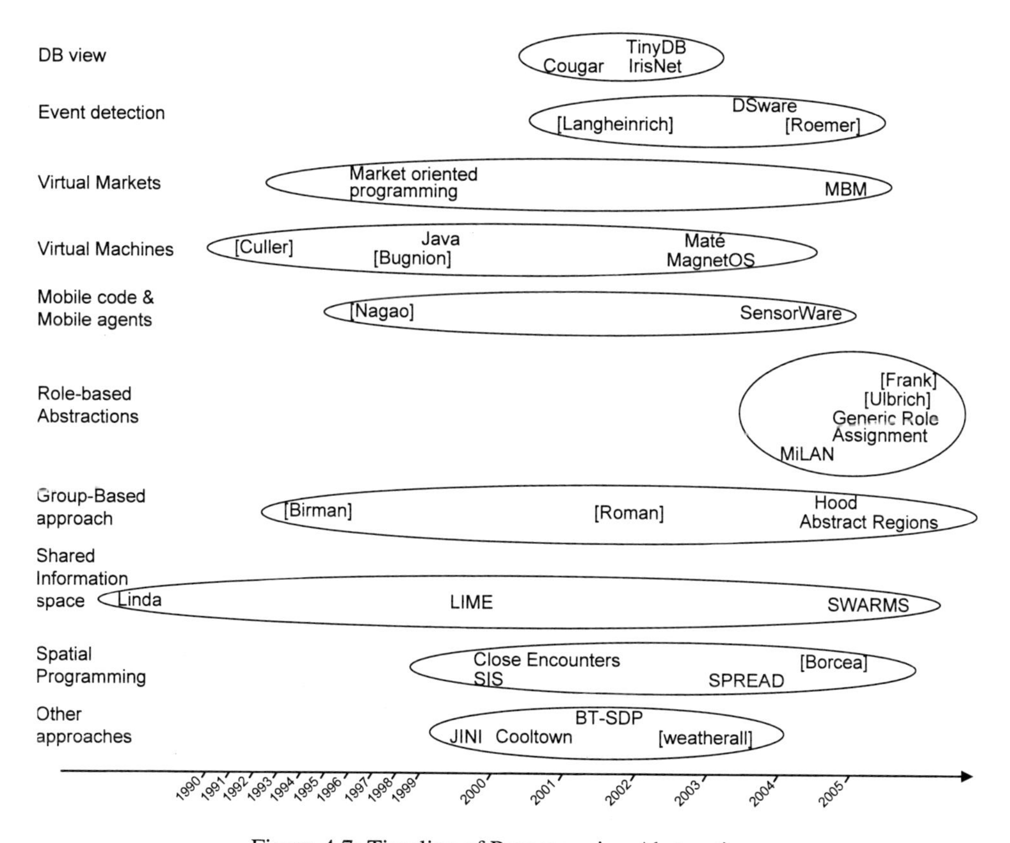

Figure 4.7: Timeline of Programming Abstractions

Support for Dynamic Code Update and Modular Composition of the Systems

Since code updates are essential for cooperating objects with long-term installations, every system will support updates in the future. For Virtual Machines, updates are most simple since the script or byte-code to interpret can be loaded from every place. For this reason, all of the VMs mentioned have this capability. Updates are more difficult for native code, since function calls and variable accessed have to be performed low-level. Either the user is willing to accept overhead for indirections, or the new code has to be fully integrated into the existing code. No general solutions exist here.

System Overhead

The overhead of virtual machines is their most limiting factor. Especially for simple commands that can be mapped to a single or only a few native commands, efficient execution methods have to be examined.

Overhead is also a limiting factor for traditional operating systems (Windows CE, Symbian) that are scaled down for use on smaller platforms.

Programming Abstraction

Programming for event-based cooperating objects have to be simplified further. The control flow of programs has to be understood easily. Multithreading or Contiki's protothreads are first steps.

Good abstractions for cooperating objects are clearly needed but they should be usable inside network of objects, too. Therefore, systems like TinyDB or Cougar are not enough.

Privacy and Security Issues

Support for relaxed and flexible security models, still ensuring some security guarantees, are needed in the context of spontaneous cooperation of objects such as PDAs. It is an important factor which limits the applicability of existing middleware.

Technological Issues

In the context of "large" cooperating objects, the energy requirements for networks of cooperating objects are still too high. IEEE 802.11 based technologies as well as Bluetooth still require too much energy to be used for continuous operation, which prevents many opportunities for real-life applications.

4.4.5 Conclusion

Most of the discussed approaches regarding the programming abstractions were designed for a specific application scenario or were tailored to some specific design goals and, thus, appear to be able to comply with only a subset of the application requirements. Therefore, for complex systems like networks of Cooperating Objects, the design a "unique" suitable programming abstraction, able to comply with the requirements and constraints of the many envisioned application scenarios appears as an extremely challenging task.

The same applies to operating system research and, although there are solutions that are well on their way to provide generic frameworks, this is something that still needs to be investigated, especially when new applications and application domains appear.

5 VISIONS FOR INNOVATIVE APPLICATIONS

Whereas the study on applications and application scenarios summarised in Section 4.1 discusses concrete applications that can be well understood and characterised today, this chapter explores visions for application areas that could potentially be realised once a wide-ranging technology of cooperating objects becomes available. The focus here is on longer-term visions that are clustered around a 10-year horizon. The document is aimed at stimulating exploratory application studies as opposed to the analytical emphasis of the other WiSeNts studies.

To achieve this goal, we elicited application visions from three sources:

1. Researchers at the WiSeNts partner institutions and our industrial partners.
2. A small scale competition held during the 2005 Dagstuhl Summer School on wireless sensor networks.
3. A large scale public Sentient Future Competition (SFC) organised by the WiSeNts personnel with substantial prizes generously donated by Deutsche Telekom Laboratories. The competition received 79 entries which were evaluated by more than 25 reviewers.

Although the design and implementation of the envisaged systems will require a lot of technical expertise and commitment, the originators of such visions need not have an immense depth of technical knowledge. Indeed there is a risk that such expertise may obscure or limit their vision. Hence we have sought to obtain contributions from as wide a community as possible.

These actions were highly successful. A wide range of visions has been elicited (a total of approximately 120) and it is our view that a substantial proportion of them merit further study. Of course any such technology-based visions inevitably have a substantial risk of encountering roadblocks in the intervening years. The roadblocks may be due to lack of technical capabilities or to social constraints.

Each application was described based on the vision timeliness and sectoral area. Timeliness refers to the research and development stages of the application, as follows:

- Immediate is used for those scenarios ready to be picked up in industrial development. As expected, a few entries (only three out of 33) were within this timescale.

- Middle-term refers to research activities that could start now but with planned development in 5 years. 18 of the entries were classified in this group.
- Long-term visions account for the cases for which any research should definitely begin after 5 years, mostly because of unavailability of the required technology and infrastructure. 12 of the visions received are believed to be of long-term.

Section 4.1 proposed application areas which can benefit from cooperating objects. Since this taxonomy addresses applications that can be understood today, we felt the need to extend this set in order to deal with some longer-term visions. We added new areas namely *human augmentation* and *enhancing social interactions*. The former refers to all the ubiquitous cooperating object technology that can be employed to assist our daily activities. The latter area uses cooperating objects to establish or maintain social relationships among people.

5.1 Overview of Visions

108 applications were received in total from the two open competitions. Out of this, we provide an overview ordered by application areas from the entries that we consider are most interesting and could potentially affect the direction of research in the next years.

5.1.1 Control and Automation

Networked embedded systems would help to enable various forms of automated and distributed process control in indoor or outdoor environments

Focus on hardware design is expected in order to realise the visions in this category. The design of specialised sensors and their miniaturisation are key technical requirements. Energy-aware data collection mechanisms should be in place. Although collaboration between sensor nodes to achieve autonomous decisions is a major challenge, it is required in some of these visions.

The **Monitoring Tape** vision is a long-term application which requires extremely small hardware in order to easily monitor the conditions of structural materials in bridges and buildings. This sensory system is comprised of sensors capable of exchanging information with other neighbouring sensor systems. The tape is easily applied to the area of interest, making construction and preventive monitoring accessible to everyone. There is a risk for using this system to spy on things or people if the sensors are tampered with to gather information on its immediate surroundings as opposed to the monitored structure itself.

The **Smarter Utilities** is a middle-term scenario that envisages the differentiation of utilities by placing low cost wireless sensors into taps, lights, switches, heating systems and appliances. Such a system should allow utilities to provide itemised billing and a whole host of value added services, well-being monitoring and remote control that might even allow the utilities to remotely enable/disable selected devices for whatever reason.

5.1.2 Home and Office

Applications in this category aim at improving the well being and space usage at home and office through the gathering of context information. The summary of section 4.1 identified applications in this category that consisted of simple sensors (light and temperature) connected through a wireless network.

Real-time low-processing signal processing algorithms are very important. This may impose high bandwidth requirements in scenario of low-resource sensor nodes. Location services including reliable and accurate localisation schemes for emergency situations. Much more effort has to be put into the information processing side, where novel techniques for information-fusion, outlier detection, and distributed calibration have to be developed. They should combine the multitude of low-quality sensory information into meaningful representation of the physical reality. These software systems need to be context aware with considerable attention to security and privacy issues.

A key aspect for the design of a future sentient computing application is providing ambient intelligence for non-expert users. Automatic, self-organising and self-managing systems will be essential for such ubiquitous environments, where billions of computers are embedded in everyday life. The middle-term **Ambient Intelligence by Collaborative Eye Tracking** provides information on both explicit and implicit subconscious social interactions and indicates directions when other communication is inappropriate. Integration of eye tracking and sentient technology will create a powerful paradigm to control and navigate applications. Ten years of progress on sensor device hardware and software should realise this paradigm, and numerous applications can be integrated into this technology.

The **Global Distributed Cooperative Sensing** envisages in the middle-term the fusion of sensors with the omnipresent portable communication devices (mobile phones, PDAs, etc.) and network enabled personal electronics (music players, digital photo cameras, camcorders, etc.) will establish a platform for a globally-distributed cooperative sensing paving the way for many novel applications at unprecedented scales. Due to the massive penetration of the above device types for the general members of the public even simple sensing capabilities will yield quite useful applications without significant impact on the price of the devices as well as the comfort and the privacy of the users.

The wealth of the produced sensory data can support endless array of applications that can simplify the date-to-date life of the general population as well as provide useful historic data that can be used for planning future public development projects, social analysis, etc.

5.1.3 Logistics

This application area include visions that strive to optimise the distribution process of goods from their production to the delivery to customers. One of them foresees a scenario where products can be traced back to its basic production even before reaching the industrial processing stage.

These sensor systems are likely to be large in scale with high mobility of the monitored goods. Distributed location and collaborative schemes need to be developed to avoid the overhead issues of a centralised back-end system. Fault tolerance is an important issue to be addressed. Further research and development on sensor design is required. In particular, MEMS/nanotechnology-based sensors that can sense chemical and meteorological conditions of the environment in which the product was produced. They should be embedded in flexible labels, similarly to the current RFID tag systems.

The **Smart Labels** vision has some aspects that can be developed today and others that require further research. The basic idea is to make value chains – including production and inbound logistics, outbound logistics, sales and marketing, as well as maintenance and recovery – smarter by using smart labels, and particularly to overcome incompatibilities at the transitions from one phase to the other. These labels should include sensors for acquiring information about their environment, together with processing and wireless communication capabilities attached to even the smallest products or parts of them, as it has already become possible with passive RFID labels by now.

The Fairtrade movement expects us to pay a slightly higher price for food and other products produced by people in third world countries to ensure that the producers receive a fair reward for their labour and investment. This would result in a more popular form of trading if the following assertions could be validated: (a) the price premium is in fact used to reward the producers of the product; (b) the producers performed a certain amount of work or invested certain assets to produce the product; (c) the product has not been contaminated. The middle-term **Validated Fairtrade** vision takes a step towards this direction. Sentient systems could track the product throughout its life, sensing or reading data about its producers, the production methods used and the environment. It adds further information to each product including the identity of the main producers with their hourly wages, work required to produce the product and also any airborne sprays detected during the production process.

5.1.4 Transportation

Visions in this category should address the safety of road users and pedestrians. Often the envisaged sensor systems would gather data for real-time or "close" to real-time information services provided by governmental agency and private organisations including insurance companies.

The visions received seem to present a different system perspective than the current transportation applications surveyed in section 4.1. Unlikely the ad-hoc and infrastructure-less characteristics, some of the visions require pre-established infrastructure of sensor nodes deployed in major roads. For instance, base stations every 1 to 5 Km and high-bandwidth backbone network. The sensor systems required vary from one scenario to another but it should include vehicle passing detector, structural material integrity, motion sensors and video capturing systems. Actuators are also discussed in a form of vibration, audible and visual (e.g.s LEDs). Most of such an information should be provided in real-time.

The **Sentient Guardian Angel** proposes the use of wireless sensor networks to address dangerous traffic situations for elderly pedestrians, children as well as for disabled persons. Communication between the networks of the participants is used to detect the threat at an early stage giving adequate warnings using suitable audible/visual actuators, alerts and instructions to the ones involved.

With the goal of improving the road traffic, the **Supportive Road** vision describes scenarios where sensors installed on the roads assist in various traffic applications including road congestion avoidance and safety of drivers. It requires significant investment in technology to be installed on roads, which might only be available in the long-term. Similarly, the long-term vision **Congestion-Free Road Traffic** takes a step further to propose a technical solution to address traffic congestion. It explores the concept of dynamic time-space corridor that can be negotiated between cooperating vehicles to guarantee congestion-free journeys from departure to arrival.

5.1.5 Environmental Monitoring

Environmental monitoring applications have crucial importance for scientific communities and society as a whole. Those applications may monitor indoor or outdoor environments. Supervised area may be thousands of square kilometres and the duration of the supervision may last years. Networked microsensors make it possible to obtain localised measurements and detailed information about natural spaces where it is not possible to do this through known methods. Not only communications but also cooperation such as statistical sampling, data aggregation are required between nodes. An environmental monitoring application may be used in either a small or a wide area for the same purposes.

Systems involved in those applications are to be infrastructure-less and very robust, and localisation plays a very important role. Since the nodes are untethered and unattended in this class of applications, the system must be power efficient and fault tolerant. Furthermore, long lifetime of the network must be preserved while the scale increases in order of tens or hundreds, and solar panel for energy should be considered. Regarding the type of sensors, the following could be considered: vehicle emissions, tree growth, other vegetation growth, wireless water sensors to monitor its quality and polluters, implantable sensors for animals, sensors for chemicals and biological components using Bio-MEMS (continuously in contact with the blood), etc.

The **LocuSent** application scenario has an immediate timeliness, although some of its technical requirements are not currently commercially available. It is a proposal for a large-scale monitor and control system for the desert locust. It should be deployed as an integrated sensor network system that can survey vast and remote areas in order to prevent outbreaks and thereby prevent the terrible famine and the disastrous economical losses that follow in the trail of the locust.

Similarly, the rather long-term vision on **Insect Monitoring and Control Networks** proposes a system that could monitor and control invasive insects in regions where they have no natural predators. These nodes would be applied to insects by spraying or alternative methods. Once attached, they would monitor the behaviour of the insect, and present corrective stimulus when disruptive behaviour was observed. Since insect behaviour is often individually chaotic but group coherent, such nodes would have to communicate with one another in order to determine the state of the individual.

Waste disposable is an urgent problem in urban environments. A more efficient and sustainable waste management system can lead to a higher life quality and less costs for the city authorities. The vision **BIN IT! The Intelligent Waste Management System** presents an RFID system that pieces of waste and encourages the correct disposal by financial incentives.

The **Zero Carbon City** measures or estimate all carbon emissions and absorption in order to charge/ration citizens according to their consumption. Individuals can receive carbon debits for their use of energy and other carbon-emitting activities and carbon credits for 'clean' energy that they generate, for example, by investing in wind farms and for carbon-absorption activities including trees and other vegetation planted or invested in. Carbon debits are converted to a tax on the individual. The direct benefits are increased environmental and public health gains. There is the risk, however, of privacy loss and fraudulent interference with sensor systems.

Water monitoring is discussed in the **Pearl Sensors** vision. Sensors are extremely small enclosed in a spherical, water-proof packaging. The goal is to monitor the water by throwing a bunch of sensors into the water and let them move with the current. The sensor pearls should be able to collect, store and communicate different types of sensor data, including location information, water temperature, water quality, stream velocity, water depth, and sound. They should also have appropriate communication capabilities to coordinate among each other as well as communicate sensor data to base stations installed in ships or on-shore.

5.1.6 Health and Fitness

Applications in this category include telemonitoring of human physiological data, tracking and monitoring of doctors and patients inside a hospital, drug administration in hospitals etc. Merging wireless sensor technology into health and medicine applications will make life much easier for doctors, disabled people and patients. They will also make diagnosis and consultancy processes faster by patient monitoring entities consisting of sensors. Those sensors will provide the same information regardless of location and automatic transitions from one network in a clinic to the other installed in patient's home will be available. As a result, high quality healthcare services will get closer to the patients. In this application localisation is important because it is critical to determine where the person is. The reliability and the minimisation of the delay between the source of the event, and the other end-point of the system is also important. The context and the person activity in the measuring time are also relevant. These application should require minimal maintenance, use of biodegradable materials, and new biosensors. The energy harvesting from the body heat seems conceivable.

The **Self-learning Children Watching** sensor network is a middle-term vision whose aim is to enable a person to sit at the playground and let their children to play. The system should notify the parents only when situations that might possibly harm his children occur. The decision if a situation might be dangerous should be taken by the sensor network autonomously based on previous experience. The benefits are likely less incidents on playgrounds. Similarly, the long-term **Small Child Care** vision creates a smart environment around a child, enabling trusted people, e.g., parents, kindergarten staff, teachers, doctors, police, to gain information about the child and to interact with him by means of his surroundings.

The rather long-term **Intelligent Pills** vision envisages the scenario that patients swallow pills that activates upon reaching the stomach. The clear benefits are the efficient use of medicine, less waste, and better dose, tailored to the patient needs. The intelligent pill takes measurements of the level of chemicals in the stomach and blood and releases the optimal quantity of medicine needed to alleviate the symptoms of the patient. The pill is built of biodegradable material and it is small enough to be expelled from the body.

The vision **A Sensor Network for our Brain** considers the implant of sensor nodes in our brain. These nodes can communicate with both external systems and the neurons. Then it could be possible to influence the storage of information and the communication in our brain without relying on our sensory system (e.g. eyes, ears etc.).

In the vision **Body Area Sensor Network for Small Children** every node in the sensor network contains a temperature and a humidity sensor, which make it possible for parents to monitor children's skin temperature not only in one point but in different points around children's body and to see if the child is sweating. This sensor network is mainly targeted for very small children who are not able to tell whether they are feeling thermally comfort. As an user interface for a system there could be for example a cell phone containing one sensor node, which would act as a sink in the network.

Finally, **Large Scale Body Sensing for Infectious Disease Control** outlines a possible future use of sensor networks for monitoring and controlling infectious diseases in large animal (and maybe also human) populations. It claims that these and other problems such as uncontrolled population growth, non-sustainable use of natural re-sources and natural disaster relief should become the agenda of future research and development.

5.1.7 Security and Surveillance

Sensors and embedded systems provide solutions for security and surveillance concerns. These kinds of applications may be established in varying environments such as deserts, forests, urban areas, etc. Communication and cooperation among networked devices increase the security of the concerned environment without human intervention. Natural disasters such as floods or earthquakes may be perceived earlier by installing networked embedded systems closer to places where these phenomena may occur. The system should respond to the changes of the environment as quick as possible. Security and surveillance require real-time monitoring technologies with high security cautions. The mediums to be observed will mostly be inaccessible by the humans all the time and hence robustness takes an important place. Furthermore, maintenance may not be possible also in these applications and then power efficiency and fault tolerance must be satisfied.

The **Human Security Network** presents the vision of a world where people can live in security and dignity free from terrorism. A world, in which when two armed teens are about to enter a high school, the weapons are instantly detected by sensor networks deployed all over the city and warnings are sent to the police and also to everyone inside the school. In this world, when you are walking late at night towards your home and you are aware of all the potential active weapons like guns, knives, and baseball bats present on different paths to home. Such information could enable you to select the safest route home.

5.1.8 Tourism and Leisure

Everybody wants to feel safe and comfortable when he is in a new environment. When you visit a new country you want to find wherever you like to go without much effort. New micro and nano technologies may help tourists in a foreign environment. For example, sensors and hand-held devices may be a city guide, or may help people in an art museum. Location of the museums, restaurants and information about weather should be provided to the tourists.

Tourism and Leisure services should be personalised. These applications must be service and context aware and cost effective. They must also support the mobility of the user.

Sensor Networks for Enhanced Human-Animal Interaction considers the communications between people and their pet animals facilitating the comprehension of animal behaviour. Advances in sensing technology will allow us to unobtrusively monitor an animal in order to translate its sounds and actions into human language. Very small networked sensors will be able to measure movement, muscular tension, body temperature, as well as capture sounds, images, and environmental conditions. These measurements will then be used to determine behaviour though a pattern recognition system.

5.1.9 Education and Training

Another emerging application area of embedded systems is the education. It is possible to provide more attractive lab and classroom activities involving cooperating objects. Current activities aim at merging embedded systems into the education methods. This application is characterised by a high dependency on the context. They all must be cost effective, affordable by many users and should have a high degree of automation.

The visions **Self-learning Children Watching** and **Body Area Sensor Network for Small Children** considered above can be also included in this section.

	Total	Immediate	Middle term	Long term
Control and Automation	7	4	2	1
Home and Office	6	-	4	2
Logistics	3	2	1	-
Transportation	3	-	2	1
Environmental Monitoring	7	1	2	4
Health and Fitness	9	1	4	4
Security and Surveillance	2	-	2	-
Tourism and Leisure	4	1	2	1
Education and Training	2	1	1	-
Human Augmentation	5	-	1	4
Enhancing Social Interaction	7	1	3	3

Table 5.1: Analysis of all the Visionary Applications. The timeliness of the visions can be *immediate:* development could start now; *middle term:* research could start now, development in 5 years; *long term:* research could start in 5 years

5.1.10 Human Augmentation

It refers to all the ubiquitous cooperating object technology that can be employed to assist our daily activities.

The vision **Agnostic Algorithms of Creation** explore the possibilities of interference among two distinct but almost identical dimensions by letting things that happen in the virtual world to reflect themselves in reality.

A Sensor Network for our Brain is a long-term vision that proposes the possibility to better influence the storage of information and the communication in our brain.

Finally, the **Father in Womb** vision considers to transport some of the mothers experiences as a pregnancy woman to the father, allowing him to follow the embryo growth, movements and sensations, providing mechanisms for interaction between both.

5.1.11 Enhancing Social Interaction

Finally, the last area uses cooperating objects to establish or maintain social relationships among people.

Personality Sensors (PerSens) is a long-term vision of a system which consists of a sensor network embedded into the clothes and other accessories, that can determine the personality type of the owner. Besides, it will also notify the owner of his behaviour in the current context and how it may appear to his counterparts.

Finally, **Ambient Intelligence by Collaborative Eye Tracking** considers the information provided by eye tracking on both explicit and implicit subconscious social interactions, and indicates directions when other communication is inappropriate.

5.2 Conclusions

This chapter discussed a number of ideas for future research. The visions classified as *middle-term* may be the basis for a research program that could start now. Table 5.1 presents some numbers regarding the visions, their application areas and timeliness. Those falling within the categories of human augmentation and enhancing social interactions pose new technical requirements including advances in wearability of tiny sensors and design of body implantable devices that are powered through energy harvesting. Cooperation is a challenge observed in scenarios for enhancing interactions among people.

6 MARKET ANALYSIS

After discussing the state of the art and the visions for Cooperating Objects, it is worth spending some time and thoughts analysing the market potential of Cooperating Objects in general and Wireless Sensor Networks in particular. Although the main purpose of this document is not to provide an in-depth analysis of current market trends, we wanted to assess the potential of the market for Cooperating Objects and get some insights from industry research departments about their use of Cooperating Object technology and what their opinion is about it.

For this purpose, we prepared and conducted a survey at one of the workshops organised by the EU called "From RFID to the Internet of Things" in Brussels in 2006. Additionally, we bought a set of market studies from ON World Inc. that cover the most important Wireless Sensor Network markets to date. To the date of publication of this roadmap, this set of studies is the most comprehensive one in the field of Wireless Sensor Networks and has, therefore, been selected for presentation here.

There are some similarities between the results obtained through our survey and the professional study performed by ON World Inc., but also differences. In the following section, we present the most relevant contents of the ON World Inc. studies that relate to the data presented in chapter 4 and then, we present the results of our own survey.

6.1 ON World Inc. Studies

The ON World Inc. study "Wireless Sensor Networks – Growing Markets, Accelerating Demands" from July 2005 is based on ongoing research since 2000 and on primary research especially for this study. The latter included phone interviews with more than 40 industrial companies, focusing on the oil and gas sector, and phone and email surveys with more than 100 OEMs, platform providers, component suppliers and system integrators.

According to this ON World Inc. study, 127 million wireless sensor network nodes are expected to be deployed in 2010. This (in their view) conservative forecast is computed by taking into account the number of deployed nodes in the following four application scenarios: *industrial monitoring and control*, *intelligent commercial buildings*, *residential control and automation*, and *advanced metering infrastructure and utility networks*. Other application scenarios subsumed under "niche markets" include agricultural and environmental control, healthcare, public safety and structural monitoring.

As depicted in Figure 6.1, from this 127 million deployed nodes in 2010, the industrial sector is the most promising and makes up for 32% of all deployed nodes, followed by intelligent commercial buildings with 28% and the residential control and automation sector with 25%.

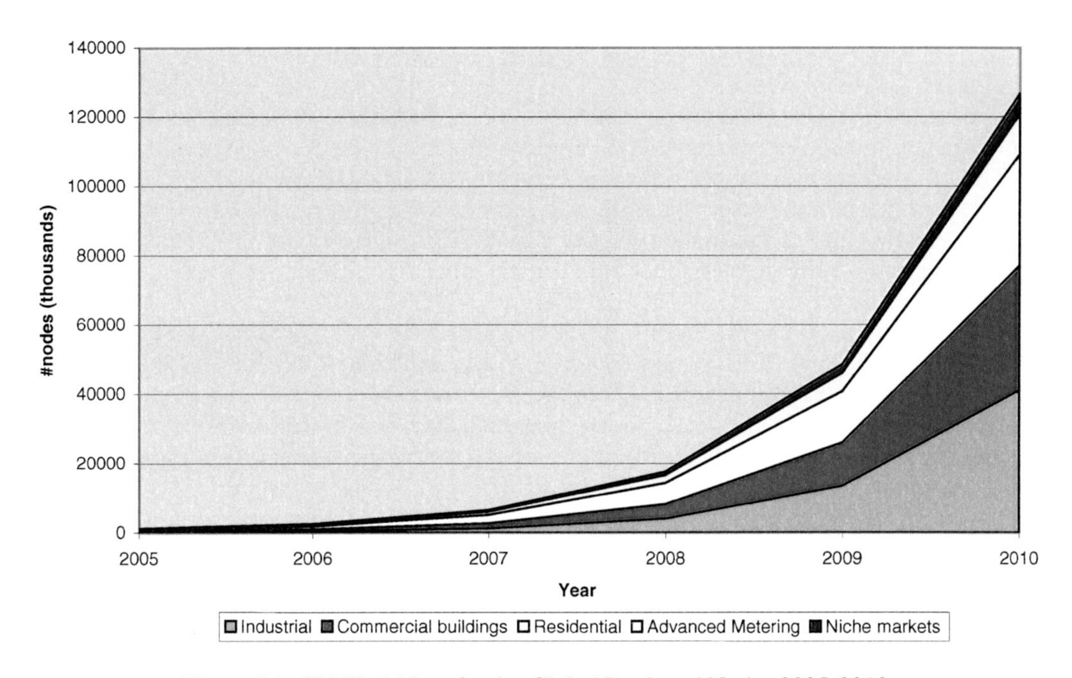

Figure 6.1: ON World Inc. Study: Global Deployed Nodes 2005-2010

As interesting as the number of deployed nodes, if not more, is the revenue acquired by the deployment of nodes. Figure 6.2 indicates that the total market is expected to be worth $8.2 billions in 2010, where 64% is attributed to the industrial sector, 17% to intelligent commercial building monitoring applications, and all other sectors produce each revenue below 10% of the total market value. However, as we will see later when we put these results in comparison to our own findings, these "niche markets" should not be neglected since the potential of quite a few of them is enormous.

Let us now look at the number of deployed nodes in each one of the areas in more detail and the reasons given by ON World Inc. as to why the development of the market is expected to be as depicted in the figures. We do not present here the graphs about the expected revenue detailed by each sector since the specific numbers and figures can be found in the studies themselves [33].

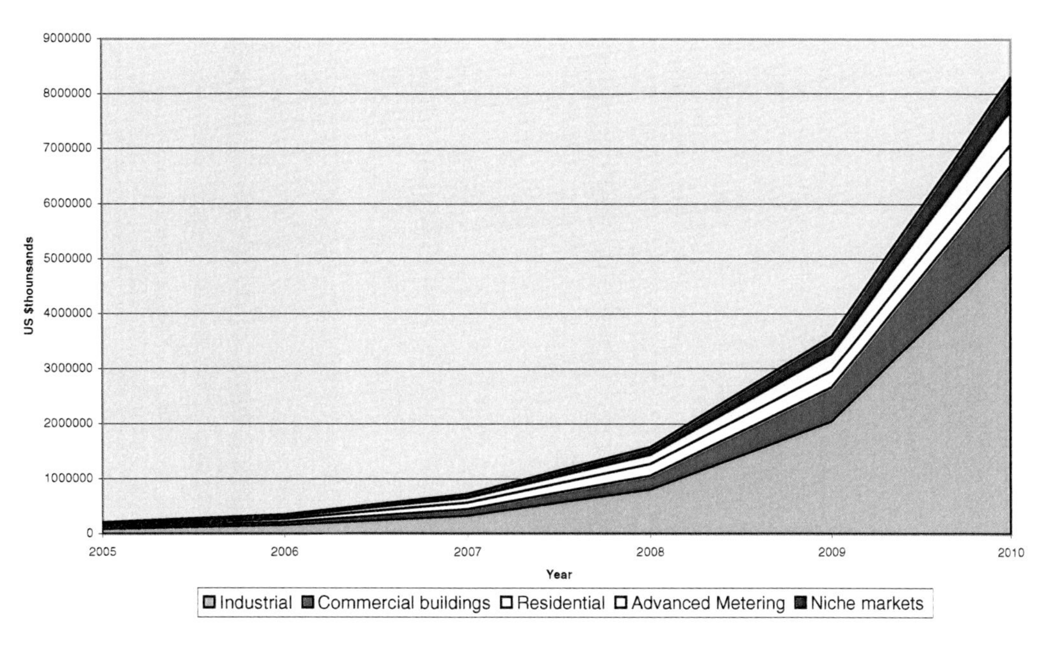

Figure 6.2: ON World Inc. Study: Global Deployed Nodes Revenues 2005-2010

6.1.1 Industrial Monitoring and Control

According to the ON World Inc. studies, this application area will be biggest in 2010, both concerning the number of deployed nodes and the amount of generated revenue. The reason given for it is an increasing need for remote monitoring of equipment since, nowadays, about 95% of all equipment is maintained manually or not at all. Especially gas and oil companies are expected to use Wireless Sensor Network technology for field monitoring and pipeline operation and they plan widespread developments within the next 2-3 years.

Figure 6.3 shows the development of the most promising application areas in the industrial sector. The monitoring of machines and other production equipment like motors, pumps and engines allows, for example, the optimisation of processes and the early detection of failures. Level, flow and pressure monitoring is used in pipeline operation and wellhead monitoring, e.g., while temperature sensor can be used for example to monitor the cool chain of goods. Tank level and leak monitoring provides both process optimisation and the detection of error conditions. Pressure Relief Valve monitoring is required in refining and (petro)chemical industries to detect failure of valves. The Health, Safety and Environmental area includes for example the quality monitoring of water and air, but also the monitoring of dangerous good that can be dangerous for water and air.

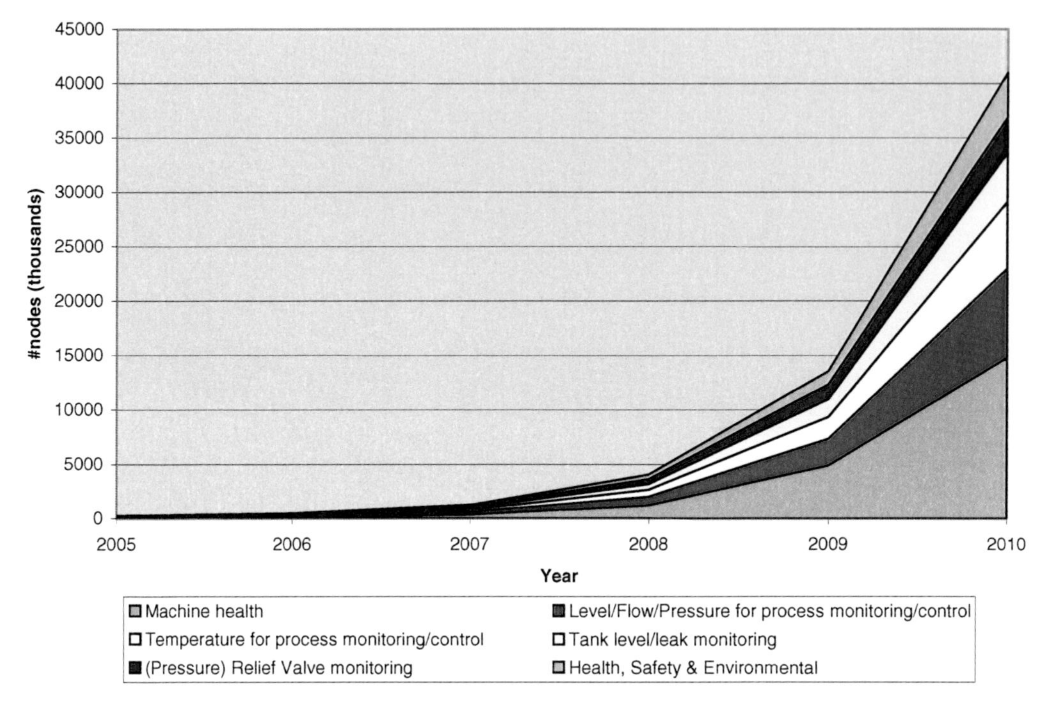

Figure 6.3: ON World Inc. Study: Global Deployed Nodes in Industrial Sector 2005-2010

6.1.2 Commercial Buildings

ON World Inc. divides the building automation sector into commercial and residential (non-commercial) since the target user and the market potential is different. These two markets differ especially in the cost per node that their users are expected to pay. For example, home users want to have much cheaper hardware and are not expected to be able to pay as much as industrial customer for each deployed node. Therefore, although the number of deployed nodes is expected to be roughly the same for 2010, commercial building revenue is likely to be 3.7 times higher than that of residential areas.

The most prominent applications are shown in figure 6.4. Heating, Venting and Air Conditioning (HVAC) applications are more comfortable if they are not centralised but can be controlled locally and are, therefore, fitted to the users. Wireless Sensor Networks also allow for the adjusting of lighting to the brightness of the environment, to the presence of people, or to other circumstances, for example emergency cases. In commercial buildings, the monitoring of gas, water and electricity usage and the possibility of easy accounting is essential. The temperature of refrigerators or deep fryers have to be monitored in commer-

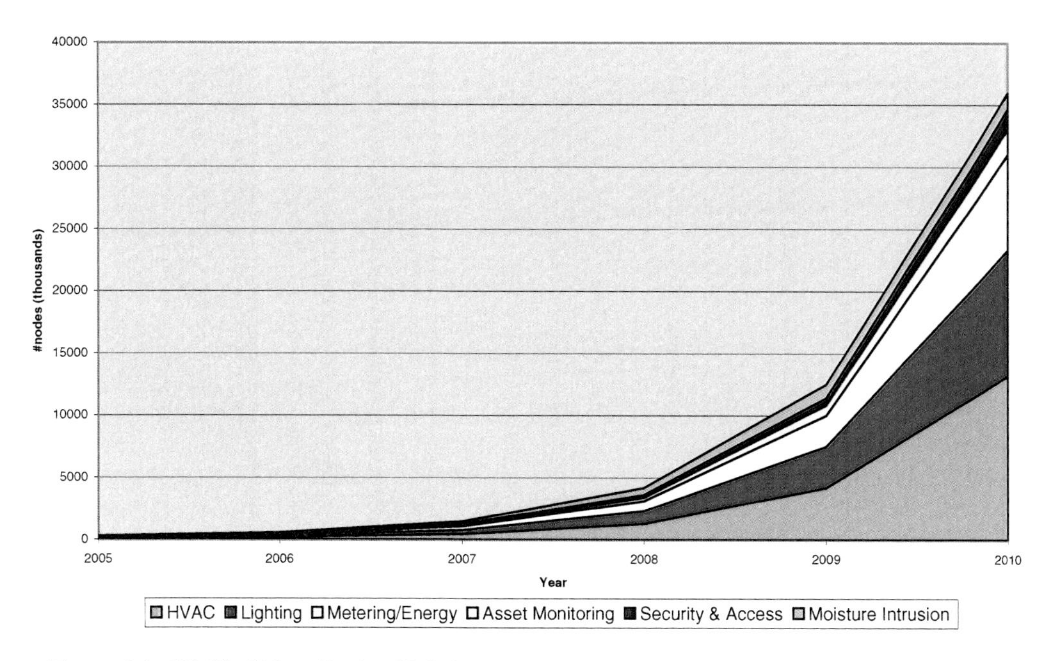

Figure 6.4: ON World Inc. Study: Global Deployed Nodes in Commercial Buildings 2005-2010

cial environments due to health regulations. Using wireless solutions, access control and security monitoring, which includes also fire, smoke or motion detection, can be deployed fine-grained. Especially for wood products, the monitoring and detection of moisture and water is interesting.

6.1.3 Residential Control and Automation

The second building control and automation sector concentrates on non-commercial, residential buildings. Therefore, the market potential is as high as for commercial buildings. Although users set other priorities, application areas are similar.

Figure 6.5 gives an overview of the most important applications for residential building applications. Lightning control and automation is already in use nowadays and will not loose its attractiveness for home users. Heating, Venting and Air Conditioning applications are not as popular as for the commercial building sector but will profit for the same reasons. Access control is not as fine-grained as for commercial buildings, but opening the garage or the front door using wireless technology can be a convenient way of applying Cooperating Objects. Also, detection of fire, smoke, gas as well as intruders by motion is desired for residential buildings. The control or automation of windows has security and convenience

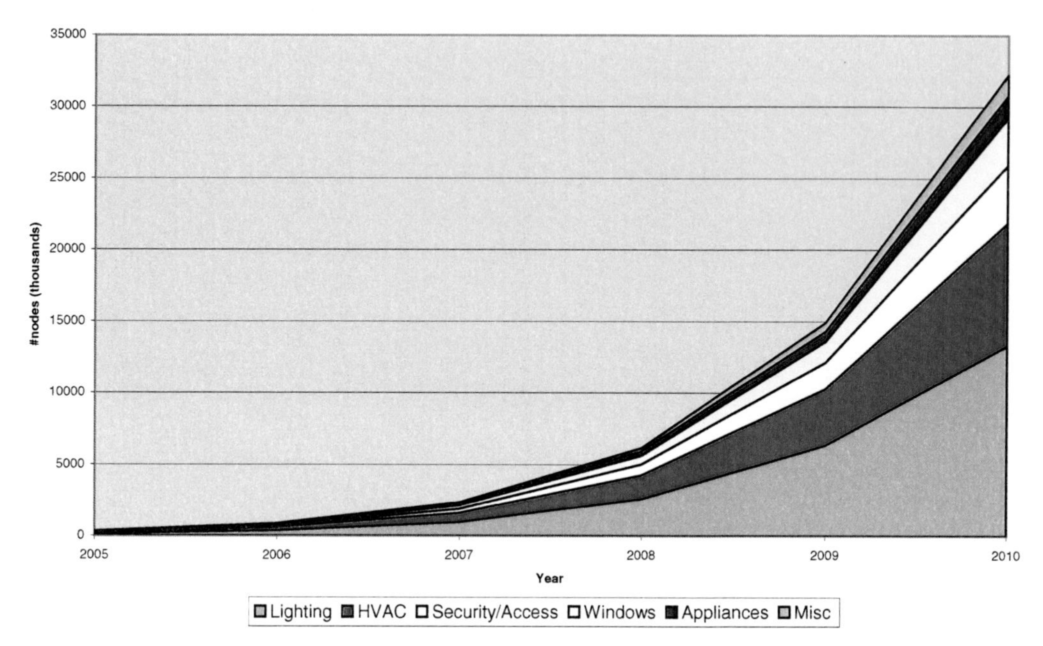

Figure 6.5: ON World Inc. Study: Global Deployed Nodes for Residential Control and Automation 2005-2010

benefits, e.g. issue warnings if they are left open or simulate occupancy. Appliances like fans, coffee makers, washers, refrigerators can also communicate with each other and other sensors in the house for example to have to coffee ready in the morning. Also, the fridge could issue a warning when items are running low. The miscellaneous sector contains for example home entertainment control which is the control of various electronic devices in a homogeneous way.

6.1.4 Advanced Automated Meter Reading

Automated Meter Reading partly overlaps with the Commercial and Residential Building area. It focuses on the reading of electric, water and gas meters for automated billing. The advanced approach includes two-way communication to enable on-demand reads and control end-point equipment for load management, for example. Gas meters could include leak detection monitoring for improved safety. Figure 6.6 shows the distribution to the different types of meters.

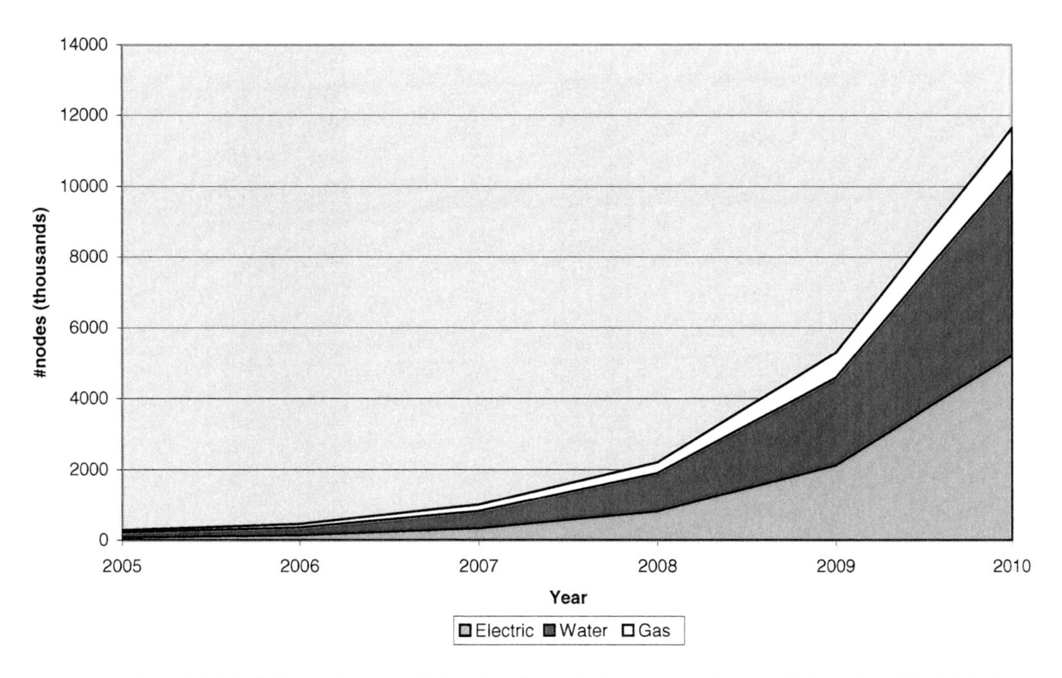

Figure 6.6: ON World Inc. Study: Global Deployed Nodes for Advanced Metering 2005-2010

6.2 WiSeNts Survey

As previously mentioned, the members of the `Embedded WiSeNts` consortium have performed a survey among the participants of the EU Workshop "From RFID to the Internet of Things" held in Brussels on March 6th/7th 2006. In it, we prepared a series of questions that were posed to volunteers that answered in an anonymous way. We obtained answers from 51 people working in both, university and industrial research and product development in the industry. The first group constituted 68% of the survey participants, whereas the latter amounted to 24%. From this last percentage, 15% of all participants worked in the logistics and transportation section and 37% in the telecommunication sector. Looking at the distribution of people from a different perspective, 27% of all answers came from participants working in software development, whereas 17% belonged to the group of hardware developers.

In this section, we present the results regarding the application domains and the acceptance of Cooperating Objects, whereas the evaluation of potential inhibitors and relevance of specific areas of research are presented in section 7.4.2.

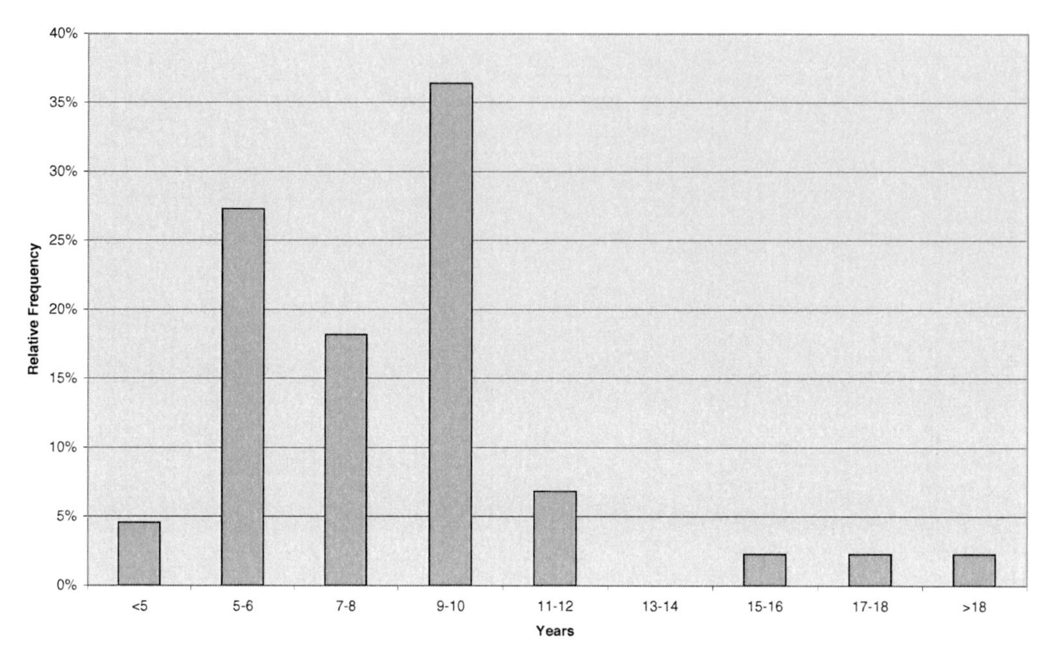

Figure 6.7: Estimation when Cooperating Objects are Used Widely in Industry

The participants were asked to give an estimation of the time they assume Cooperating Objects will be widely used in the industry. The results, presented as a relative frequency in Figure 6.7, coincides with the results found in the ON World Inc. studies. It is interesting to notice that over 80% of all participants estimate that the breakthrough for the technology will happen in the next 5 to 10 years and it is safe to assume that they assume that most technological inhibitors and research issues will be solved by then.

We have taken this into consideration when creating the timeline of research in the next chapter and, interestingly enough, coincides with the view of university researchers and industry research institutions regarding the point in time of major breakthroughs that will allow for the use of Cooperating Object technology on a wider basis.

As expected, the relative frequencies of Figure 6.7 also indicates that there is a certain number of optimists that believe Cooperating Object technology is almost ready for prime-time, and also a slightly larger number of pessimists (or realists?) that do not expect Cooperating Objects to be widely used in the industry until at least 15 years from now.

Figure 6.8 shows the results of the survey regarding the significance of Cooperating Objects for all application scenarios given in section 4.1. The x-axis contains a scale from 1 to 5, where 1 indicates not significant application domains and 5 highly significant. We asked the participants to rate the relevance of the application domains as they expect them to be in the short term (0 to 5 years and depicted in blue in the graph), medium term (5 to 10 years and depicted in purple in the graph) and in the long term (10 to 15 years and depicted in yellow).

According to the figure, Cooperating Objects seem to be very significant in the medium to long term for all applications, although application domains such as *Tourism* and *Education and Training* seem to be considered less relevant in general than other domains such as *Transportation* and *Logistics*.

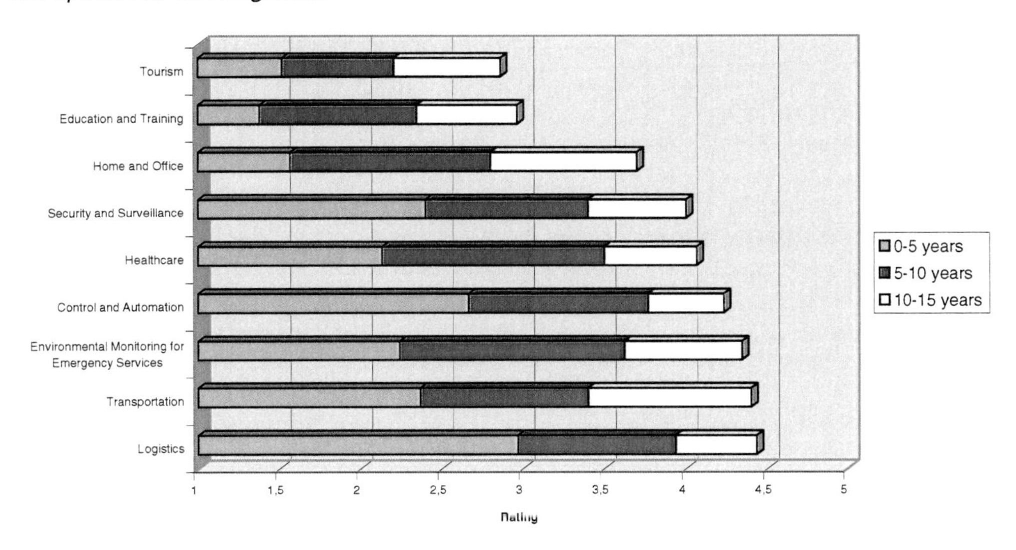

Figure 6.8: Significance of Cooperating Objects for Applications

On the other hand, application domains such as *Home and Office* have a high potential of becoming relevant in the medium and long term, and they do not seem to be there yet in the short term to be considered a key application domain. On the other hand, domains such as *Logistics*, *Transportation*, *Environmental Monitoring for Emergency Services* and *Control and Automation* are considered highly relevant in the short term and will probably remain highly relevant in the medium and long term, making them good candidates for industrial products. The same holds for the *Security and Surveillance* and the *Healthcare* domains, although the absolute values obtained by our survey indicate a slightly lower relevance. This could also be related to the higher relative numbers of participants of the survey coming from both the logistics and transportation product development areas.

However, it is interesting to notice that the interpretation of our results and those of the ON World Inc. studies are similar in nature and seem to mirror the views of both the research communities and the industry.

6.3 Conclusion

As mentioned at the beginning of this chapter, the main purpose of this roadmap is not to provide an in-depth market study of Cooperating Object research, but we wanted to forge our own picture of what Cooperating Objects could achieve in the market.

It is interesting to see that, although the application domains of the ON World Inc. studies and our own application domains do not entirely match, the results provided by our survey could be mapped to the findings of ON World Inc. very well. Additionally, as we will mention in the next chapters that deal with the timeline of research in the field of Cooperating Objects, the times provided by the industry seems to be inline with the assumptions and timelines of research. Of course, this does not mean that both the research community and the industry could not be wrong, but it is reassuring to notice similarities in their views.

In order to interpret both the results of our survey and the ON World Inc. studies, the application domains can be mapped as follows: The *Industrial Monitoring and Control* domain can be mainly mapped to the *Control and Automation* and (partly) to the *Environmental Monitoring for Emergency Services* domains of section 4.1. The *Commercial and Residential Building* sectors can be mapped to *Home and Office* and partly to *Security and Surveillance.* Finally, the *Advanced Automated Meter Reading* sector is fully covered by the *Home and Office* domain. All of the "niche markets" of the ON World Inc. studies can be mapped without losses to the other application domains of section 4.1.

In a more direct comparison, both the study and the survey predict a high relevance for *Industrial Monitoring and Control* (ON World Inc.) and *Control and Automation* (`Embedded WiSeNts`). In contrast, it is interesting to notice that the *Home and Office* application area has received a relatively low short-term relevance rating in our survey, but has a high potential according to the ON World Inc. studies. The other way round, *Healthcare* is only a niche market for the ON World Inc. studies, whereas it is one of the middle-term solutions the participants of our survey expect to appear as part of Cooperating Object research and commercialisation. Interestingly enough, two of the most relevant application domains in our survey (*Transportation* and *Logistics*) do not belong to any of the main sectors in the ON World Inc. studies.

The reason for this apparent discrepancy lies in the way relevance is assessed in the two studies. In the case of ON World Inc., relevance is proportional to the estimated number of deployed nodes, whereas in our case, we assess the significance of Cooperating Objects for specific domains based on an abstract scale. In our opinion, even if a market analysis might indicate that the number of deployed nodes in these areas is not high, there seems to be a market for it in terms of the expected areas of deployment.

7 RESEARCH ROADMAP

After exploring the state of the art in terms of research activities, future visions of application of Cooperating Objects, and their expected potential regarding deployments and possible revenue of different sectors that will benefit from them, let us now investigate in more detail the work that still needs to be done at the research front.

In this chapter, we give an insight about the current trends and gaps that need to be considered for future research. We put this research in perspective for the next 10 to 15 years in section 7.2, where we discuss what we think is a realistic time scale for research in each of the identified areas, as well as the point in time where we expect to have a major technological breakthrough. Finally, in section 7.3, we give an estimate of the evolution of the market based on the professional studies from ON World Inc. and on our own insights regarding the evolution of research.

7.1 Gaps and Current Trends

Armed with information obtained from chapters 4 and 5, we now turn to the description and explanation of the relevant research gaps for Cooperating Object research. These gaps have been grouped into 5 different categories as follows:

Hardware: These are gaps that have to do with the development of the devices that physically constitute networks of Cooperating Objects. The gaps that fall into this category are: *Sensor Calibration*, *Power Efficiency*, *Energy Harvesting*, *New Sensor and Low-Cost Devices* and *Miniaturisation*.

Algorithms: These are gaps that deal with functional properties of Cooperating Objects, that is, specific protocols, types of procedures, etc. The gaps that fall into this category are: *Localisation*, *Context-aware MAC and Routing*, *Clustering Techniques*, *Data Storage and Search* and *Motion Planning*.

Non-functional Properties: These are gaps that deal with Quality-of-Service-type characteristics. Properties that do not affect the functionality of the network, but its quality. The gaps that fall into this category are: *Multiple Sinks*, *Scalability*, *Quality of Service*, *Robustness*, *Mobility*, *Security*, *Heterogeneity* and *Real-time*.

Systems: These are gaps that have to do with the specific architecture or support for the rest of the system. Normally, systems work at the individual Cooperating Object level, but have to provide support for networking. The gaps that fall into this category are: *Adaptive Systems*, *Operating Systems*, *Programming Models* and *System Integration*.

Others: This category collects the gaps that do not fit anywhere else or that might be hard to classify within other categories because they do not really fall into the computer science umbrella. The ones we have selected for description here are the following: *Modeling and Analysis*, *Experimentation* and *Social Issues*.

This same classification will be used through this chapter and for the presentation of the timelines in section 7.2. Let us now describe each of the gaps in detail.

7.1.1 Sensor Calibration

Description and Relevance

Low cost is a key requirement for the widespread adoption of sensor networks for real-world applications. Hence, sensor nodes typically employ low-cost, uncalibrated sensors rather than expensive, factory-calibrated sensors. Such low-cost sensors are often very unstable in that their output depends on aspects like supply voltage and environmental conditions (e.g., temperature). In addition, sensor output is often affected by sensor orientation or dirt.

In such a setting, it is a difficult task to ensure that the output of a sensor network really mirrors the ground truth of the real world. This is exactly what calibration tries to address. There are two issues to consider here. First, it must be ensured that sensor output is consistent across the network, i.e., that two different sensors output the same value if they are presented the same physical stimulus. This is also called relative calibration. Secondly, we have to make sure that sensor output matches the real world, i.e., if the ambient temperature is 25 degrees Celsius, then a temperature sensor should also report the equivalent of 25 degrees Celsius. This is also called absolute calibration.

Relative calibration can exploit the fact that some physical quantities change only slowly over distance (e.g., temperature), such that sensors which are close together should report similar values. Absolute calibration is typically achieved by temporarily installing a second, calibrated measurement device alongside the sensor network to measure ground truth. For example, many installations use a set of video cameras to observe what is happening in the network and to calibrate the output of the actual sensor network. Additionally, other settings use mobile nodes with appropriate sensors to perform the calibration of static nodes.

Existing Trends

While some approaches for calibration have been proposed in the recent past, actual calibration solutions are often ad-hoc and require a large amount of application-specific engineering. In many cases, the calibration infrastructure is at least as complex as the sensor network itself. Significant work is needed to arrive at a systematic treatment of calibration issues. Also, ready-to-use tools are needed to support calibration in practical settings.

7.1.2 Power Efficiency

Description and Relevance

Power has been one of the most important issues for the electronic world since the very beginning. The modern world has almost no problem with supplying power for fixed systems, like servers or home PCs, but mobile systems are a different issue. Usually, batteries are used that need to be periodically recharged. Since this is infeasible in some scenarios, many new technologies have been created to feed these systems with power – see gap *Energy harvesting* in section 7.1.3.

Power efficiency should, therefore, be examined from two different points of view: power-efficient hardware and power-efficient software. Current battery technology is able to store enough power for long hours of discontinued operation, but their capacity is limited by their size. Unfortunately, battery size has to be properly scaled to the size of the device it is used on and, therefore, smaller devices do not have the batteries yet that would allow them to achieve the lifetime needed for their continuous operation. On the other hand, power-efficient algorithms are needed to be able to make appropriate use of the available power and not to drain the battery without a real need.

Cooperating Object systems require the cooperation of several (possibly mobile) devices that work together to perform a common task. However, the individual components of such systems are very small, and so are their batteries. Connecting them to a wired power supply is not possible and the use of alternative power sources such as the sun (solar cells) is not always a solution. Additionally, these devices are required to adapt to their environment and not to rely on specific power sources such as solar energy.

Moreover, Cooperating Objects communicate with each other using wireless technology and, depending on the deployment, even communicate continuously over extended periods of time. Therefore, the need for power-efficiency hardware devices and power-efficient algorithms are two of the most important gaps related to Cooperating Object research.

Existing Trends

Regarding hardware, low energy processors and controllers have been designed and used, especially in the area of embedded controllers and devices. There are also advances in simple low lower sensors, e.g. for temperature or humidity, while other areas are not showing similar advances, e.g. gas or air movement. However, the more efficient a particular piece of hardware is regarding energy consumption, the more expensive it becomes. Unfortunately, cost is a definite constraint in the Cooperating Object area and, depending on the application scenario, people would not be willing to pay too much for each device. Therefore, there is a need for low-cost, power-efficient hardware.

Regarding power-efficient software and algorithms, the research community is currently working on individual solutions for specific problems. For example, an application might be able to switch off certain parts of the hardware if it knows it will not need it in the next time. Using this technology, the radio interface could be switched off if it is not used for an extended period of time. In other approaches, data is aggregated within a cluster and then sent to the cluster head, that takes care of sending it to the appropriate recipient. However, most of these solutions only work in certain conditions, for specific types of algorithms, etc. and lack a generic solution that can be used in a wide variety of application domains.

7.1.3 Energy Harvesting

Description and Relevance

The main idea behind energy harvesting is to leverage the use of alternative forms of available energy within the environment in order to transform it into electrical energy that can be used to power a device partially or totally. In most cases, solar, mechanical or thermal energy is used as an alternative source.

Depending on the application domain, Cooperating Objects could be equipped with solar panels that power the network. Since the efficiency of inexpensive crystalline silicon solar cells is low, they usually have to be exposed to bright sunlight or the illuminated area has to be large. Additionally, solar cells are relatively big in comparison to the sizes of devices in the network, so that in some scenarios, solar energy is not an option. Therefore, there is a need for more efficient solar cells that produce more power with less photovoltaic excitation, even when the exposed surface is comparatively small.

Other approaches based on thermoelectric elements can convert temperature differences into electric energy. The higher the difference, the more efficient is the thermoelectric element. Although the use of such thermal elements only produce a few μW from human body temperature and, therefore, cannot be used well into home and office environments, this equipment could be used in more industrial application domains where, for example, hot liquids need to be monitored.

An alternative to using thermal energy is to use the oscillation of a proof mass to collect energy from vibrating parts. Again, smaller generators produce only a few μW, whereas much bigger systems are even able to produce several mW. Therefore, these elements are not suitable for all scenarios, especially when hundreds of small Cooperating Objects are to be deployed.

Finally, strain on piezoelectric material results in electrical power in the order of several μW. As with the previous technologies, the energy obtained from such systems is not enough to power complete devices directly and, therefore, it is necessary to either combine several technologies or to use energy harvesting techniques to recharge more traditional batteries. However, both would need to be developed.

Existing Trends

Passive RFID tags are powered from a high-frequency electromagnetic field generated by the RFID reader. The RFID tag is charging a capacitor that provides enough energy to send back the data stored on the tag. Current research at Intel extends a RFID tag with sensing technology: the sensor data is read out using the normal RFID reader.

Existing solutions for home automation manage to harvest enough energy from the act of pressing a light switch to send on/off commands to the light. Since the light switch has no function during the remaining time, it does not need a battery to power a radio in order to receive messages.

Current research in the field of energy harvesting tries to combine existing techniques to create more efficient power generators, although there is definitely the need to improve the energy generation capabilities of individual techniques. New materials, such as electroactive polymers, are being examined since they promise a higher energy conversion coefficient.

7.1.4 New Sensors and Low-cost Devices

Description and Relevance

Wireless sensors are among the most widely used devices for Cooperating Object applications. Recent advances in the microelectromechanical area have made it possible to produce low-size, low-power, low-cost sensors. However, considering some real-world applications, sensor size is not as cheap as targeted yet. Wireless Sensor Network applications may consist of hundreds to thousands of sensor nodes and, as a consequence, low-cost sensor node design is a must.

The development of Micro-Electro-Mechanical Sensors (MEMS) has led to the production of low-cost sensors. MEMSs provide not only low-cost but also low-powered, low-size sensor nodes. Also production in large volume significantly reduces the cost. As Cooperating Object application areas spread, the need for low-cost sensors increases. Some applications such as environmental monitoring, surveillance, and disaster relief may need to use vast numbers of sensor devices. New efficient and low energy imaging sensors with significant processing capabilities are needed in these applications. Also some sensors in the network may fail due to the environmental conditions and energy constraint that requires the deployment of redundant number of nodes. In such cases, low-cost sensor is a necessity. Otherwise, the implementation of such applications would be hard or even impossible.

Existing Trends

There are quite a few sensor device manufacturers in the world that try to reduce the price of sensors in order to make them viable for large-scale deployments. However, the design of new devices is not a cheap undertaking and the lack of standard interfaces make it even harder for newcomers to enter the market.

Nowadays, the typical sensor node price lies between \$50 and \$200. On the other hand, applications requiring more than 100 sensor nodes dramatically increase investment costs. Silicon-based tilt sensors offer cost effective solutions over older fluid-vial sensors. The ultimate target is to produce sensor nodes with a price of under \$1.

7.1.5 Miniaturisation

Description and Relevance

Research in the field of Cooperating Objects started with the idea of providing enough computing resources to our everyday lives and to make them more comfortable by the use of computing technologies. These systems are expected to be used in the most varied environments: from inaccessible areas to our own offices or even our own bodies.

Therefore, the sheer numbers of devices that we will have to share our lives with imply that size, biodegradability, etc. play a very important role. Most embedded devices and, of course, sensors too, are designed to be near the source of information they are supposed to monitor. The need for miniaturisation arises then by the need to pollute the space as little as possible and to interfere minimally with the observed phenomenon. Additionally, the smaller the devices, the easier it is to carry them around with us or to embed them into the environment seamlessly.

Existing Trends

Wireless sensor network devices available today on the market can be considered very small-sized computers. An average sized node in research is around 13 cm^3 (Mica2) or slightly bigger, a small sized mote is around 2.9 cm^3 (Mica2Dot). Despite their small size when compared to traditional systems, they are still too big to be embedded in small objects of daily life, e.g. in smart home scenarios. System on Chip solutions with smaller dimensions exist, but are usually tailored to specific scenarios. Moreover, the key problem is often the battery and packaging and not the sensor node itself.

The vision of Smart Dust is to be able to implement computing devices the size of a grain of sand that will contain sensors, computational power and be able to communicate wirelessly with other devices. Although we are still very far from the realization of this vision, there are already devices available that measure less than 1 cubic millimetre. But there is still the need for small hardware, at an acceptable cost, with enough computational power to implement the vision of smart dust.

7.1.6 Localisation

Definition and Relevance

Provisioning of localisation information in distributed networks has been, and still is, an interesting research topic. Indeed, localisation can be both a service offered by the network to the applications (object tracking, mapping, automatic driving, and so on) and a technique for improvement of the network operation itself. For instance, many algorithms for Wireless Sensor Networks assume the availability of location information on the nodes. Some algorithms assume only that each node is acquainted (in some way) with its own spatial coordinates, others require the knowledge of the positions of the surrounding nodes only or of all the nodes in the network. Algorithms that make use of localisation information for managing the medium access or the routing are usually much more efficient than location-independent algorithms, though they may turn out to be excessively sensitive to localisation errors.

Many localisation methods have been proposed, including the distance estimation based on the strength of the radio signals received by surrounding nodes, measurement of the time of flight of light and pressure impulses, broadcasting of geographical coordinates of beacon nodes, and so on. Unfortunately, despite the great interest that this topic has focused in the research community, the solutions proposed in the literature often reveal their limits when realized in practice. Indeed, first experiments in this direction have revealed that signal strength measurement does not allow a fine localisation, even after long and precise calibration. This is due essentially to the unpredictability of the radio channel dynamic. Furthermore, energy level has impact on the transmission power of a node, so that the calibration process becomes rapidly loose when nodes progressively discharge their batteries.

Therefore, the topic requires further investigation, in particular through experimental campaigns in various environmental conditions. Also, the sensitivity of the location-based algorithm to the tolerance in the location information has not been sufficiently covered in the literature yet and deserves further investigation.

Existing Trends

Recently, some localisation mechanisms based on the comparison of the time of flight of impulses with different propagation speeds have been proposed. However, such systems usually require either cumbersome equipment, are not very energy-efficient or else, prone to errors due to environmental noise. Nonetheless, some applications based on such a paradigm have been recently commercialised.

Another interesting trend in this field encompasses the use of Ultra Wide Band (UWB) technology for ranging. Indeed, the ultra-wide frequency spectrum of UWB impulses permits a very fine time-resolution at the receiver, which can be exploited to estimate the distance between transmitter and receiver. Some first commercial products like Ubisense are available. However, concerning the integration into small Cooperating Objects the technology is still in a development phase.

7.1.7 Context-aware MAC and Routing

Description and Relevance

Generally speaking, a context-aware system can be defined as a system that can extract, interpret and use context information and adapt its functionality to the current context of use. The term context has a rather broad meaning. A possible definition proposed in [17] is the following:

> Context is any information that can be used to characterise the situation of an entity. An entity is a person, place, or object that is considered relevant to the interaction between a user and an application, including the user and application themselves.

Until now, the context-aware paradigm has been mainly applied to the application layer as, for instance, in the design of office and meeting tools. In that framework, however, the *context* information shrinks to a limited set of indexes, such as location, time of the day, date, user profile, and hardware settings (e.g., device's screen size and definition).

However, in the framework of Object-based systems, the concept of context-awareness has to be extended, at least, in two directions. First, the term *context* has to be enlarged, in order to include heterogeneous environmental information, such as node density, mobility pattern, hardware constraints and capabilities (energy, transmission rate, coverage) and so on. Second, the context-aware paradigm has to be applied to other protocols in the software stack. For instance, in the scenario of inter-vehicular communications, the definition of MAC and routing algorithms that can adapt their parameters to the context, defined by vehicles speed, cars density, circuit typology (urban, rural, highway) etc, can lead to relevant performance gain with respect to solutions proposed for generic ad-hoc networks.

Nonetheless, the application of the context-aware paradigm to the design of MAC and routing protocols is still in its early stage. The challenge for such systems lies in the complexity of capturing, representing and processing contextual data. In addition to being able to obtain the context-information, systems must be able to process the information and to deduce the meaning. Indeed, context is often determined by combining different pieces of limited information independently produced by several sources (sensors, RFID, user profile info, etc). This is especially important in mobile computing where the context changes frequently and rapidly.

Existing Trends

The Cognitive-radio paradigm probably represents one of the most relevant trend concerning context-aware protocols. Cognitive radio was initially thought of as an extension of the software-defined radio concept (Full Cognitive Radio). The goal was to develop systems able to change particular transmission or reception parameter to fulfill specific tasks. From this initial vision, however, the research has been progressively specialised on the utilisation of TV bands for communication. Hence, currently the essential problem addressed in Cognitive Radio research is the design of high quality spectrum sensing devices and algorithms for exchanging spectrum sensing data between nodes.

Another relevant trend concerns the context-aware MAC and routing algorithms for dissemination and retrieval of information in a dynamic heterogeneous network, where intelligent devices (cellular phones, PDAs, laptops), coexist with very simple devices (sensors, RFID tags), and with an underlying network of sensors and storage elements. In this scenario, the processing and storage capacities, as well as the wireless communication capabilities are widely different among nodes. The design of context-aware mechanisms for gathering, processing and delivering data across the network is, hence, one of the most promising and challenging problems that have been recently considered in the research community [21].

7.1.8 Clustering Techniques

Description and Relevance

Wireless Sensor Networks are deployed in large number of unattended nodes and hence the underlying network architecture has become one of the challenging areas in wireless sensor network research. A common network architecture of deploying the sensor nodes is to employ network clustering. Such cluster-based architecture assigns for each cluster a cluster head that can usually reach all the nodes in the cluster in one hop. Also, each cluster head is responsible for the interaction and collaboration with other cluster heads in the network, in addition to the interaction with the network base node, to perform the required tasks. Furthermore, applications requiring efficient data aggregation are natural candidates for clustering.

Clustering helps the nodes to minimise the overall energy dissipation in the network by allowing only some nodes to take part in the transmission to the base station. Moreover it also helps to reuse the bandwidth and thus utilises better resource allocation and improved power control. Furthermore, for a cluster-based sensor network, the cluster formation plays a key factor to the cost reduction, where cost refers to the expense of setup and maintenance of the sensor networks. Clustering can be extremely effective in one-many, many-to-one, one-to-any, or one-to-all communication. For instance, in many-to-one, clustering can support data aggregation and reduce communication interference.

Appropriate cluster-head election can drastically reduce the energy consumption and enhance the lifetime of the network. Several policies have been developed in recent years for cluster-head election, but there are still open issues to be addressed. For example, in some approaches each node probabilistically decides whether or not to become the cluster-head by using local information. However, there might be cases when two cluster-heads are selected in close vicinity of each other. Moreover, the node selected can be located near the edges of the network, in which the other nodes will expend more energy to transmit data to that cluster-head.

Clusters are formed by considering many factors such as communication range, number and type of sensor nodes and geographical location, that can be based on GPS or other techniques. However, the cluster formation problem is not usually considered in general purpose clustering methods, whereas research is more focused on the issue of network management within the cluster, particularly energy-aware routing. Therefore, another aspect that requires further research is the cluster formation policy.

In wireless sensor networks with mobile nodes, it is not necessary (relying on direct or multi-hop connections) to transmit information from the static nodes to the sink. Indeed, mobile nodes can relay the information among the static nodes and the sinks, also in absence of permanent wireless connectivity among them. Therefore, the static nodes could be also organised in clusters and the mobile nodes could visit them and transmit the information to others sinks or to the main sink allowing energy savings in the static nodes. Ideas from existing research in clustering techniques could be applied and adapted to the particular requirements of this approach (location of the nodes, delay problems , etc). Therefore, it is a promising area where significant research and experimentation is still required. .

Furthermore, when several nodes in the Wireless Sensor Network are mobile, modification of the cluster structure in the presence of topology changes leads to performance degradations in the network. Therefore, further research in clustering maintenance is required to keep the cluster infrastructure stable in the face of topology changes.

Existing Trends

Many wireless network protocols and clustering algorithms have been developed in recent years for increasing energy efficiency. Existing work is mainly focused on energy-aware routing and many successful approaches, such as LEACH, have been developed. A novel application of clustering in the context of Wireless Sensor Networks with mobile nodes has been recently considered. This application has particular characteristics that have not been tackled yet, requiring further research in the next years. Among others, adaptation to topology changes, dealing with delay problems and the use of geographical location information in the algorithms could be mentioned.

7.1.9 Data Storage and Search

Description and Relevance

In many application areas, large amounts of sensor data are collected by a reasonable number of tiny nodes in order to assist users in the decision making process, which may trigger some actuation in the monitored area.

While collecting sensor data it is crucial to minimise communication cost and delay for the application tasks. To address this, efficient storage and querying of sensor data are both critical and challenging issues in systems of cooperating objects.

Although current research efforts focus on the design of storage and search systems in homogeneous and stationary wireless sensor networks, research is still needed on:

Consistent filtering and aggregation methods. Efficient data storage and search mechanisms are needed to bring data together in a useful and consistent form. To achieve this, data fusion using filtering techniques and query-based or model-based data aggregation are key approaches. The results of such data processing tasks can be stored with certainty probabilities. The usefulness of fused data (yielding higher probabilities) is highly application dependent.

QoS models for sensor data. A mobile sensor system architecture and real-time data delivery seems inevitable, but the question is with what QoS (latency, reliability). QoS mechanisms are needed that assign cost and value to timely sensor data (including aggregated data).

Strategies to exploit geographic locality. The sensor data of interest to a given Cooperating Object is most likely to originate from a nearby sensor. Also, the observation areas of cooperating objects often overlap. Geographic storage can be based on imprecise location (e.g. maybe based on grid squares). Mobility complicates this issue further as arbitrary routing between two nodes (i.e., direct network connectivity) may not be present. So cooperating objects should explore geographic locality and network proximity to share data and cooperate.

Communication and energy optimisation . Distributed storage and search strategies are needed that are optimised for communication cost and energy utilisation. Cooperating objects may be limited in both memory and computational resources. Thus, they may be unable to buffer a large volume of sensor data for a long period of time.

Robust storage and search algorithms. Storage and search mechanisms are needed for mobile scenarios that are robust when connectivity is intermittent. The type of mobility (low and high) creates a set of unique system design issues with the high mobility scenarios being the most difficult ones to deal with (e.g. inter-vehicular networks (IVNs)) because of rapid changes in network topology

Integration of heterogeneous sensor and Cooperating Object systems. Data storage, retrieval and search schemes that cope with various types of data gathered from heterogeneous sensor systems. Heterogeneity here refers to different types and manufacturers of sensor and actuator devices. Each of those can generate large amounts of sensor data that need to be adequately represented before it can be stored.

Support for calibration and fault detection of available sensors. The calibration of Cooperating Object systems requires data redundancy, which can be achieved by replicating sensors/actuators and/or replicating Cooperating Objects in a given area. Calibration and fault detection is more difficult for mobile systems since sensors are moving in space and time.

Existing Trends

There is a considerable research interest to develop efficient data querying schemes. Such approaches, however, are designed specifically for networks that have fixed homogeneous sensor nodes and are based on the assumption that all nodes try to convey data to a central node (often called sink node).

Most large-scale applications exhibit strong geographic locality. That is, the sensor data of interest to a given Cooperating Object is most likely to originate from a nearby sensor. An example is the monitoring of a duct valve of a high pressurised fluid. The actuation action, which can be regulating this valve, is likely to need the pressure readings from the nearby sensors.

Current storage and search strategies for large scale system benefit from the use of data models that exploit locality without loss of generality. For instance, data-centric or location-aware search and data dissemination schemes where intermediate nodes can route data according to its content or the location of the nodes have been employed in these approaches.

In this direction, the trend is to perceive a network of sensor nodes as a distributed database that may hold data locally for certain types of user's query (distributed storage). The approaches put forwarded consider possible resource limitations of cooperating objects such as tiny memory, small processing unit and limited energy available. Thus, resource optimisation, scalability and power consumption are practical problems that have been studied.

To consume less power, the trend is to employ data aggregation techniques in order to reduce the number of data packets conveyed through the network. However, there is a trade-off between the amount of data that can be held locally within the cooperating object and the extra data that should be forwarded to a data collector. This makes any strategy for local data persistence a real challenge since communication costs are high in wireless networks.

Greedy flooding and its variant with randomisation (gossiping) are the most used techniques in naive dissemination protocols proposed for Wireless Sensor Networks. To overcome the high cost of message exchanges, other alternatives have been proposed.

In some protocols, the dissemination process is stimulated by sensor nodes with the broadcasting of advertisement packets (SPIN approach). In others the sink gathers data from the sensor nodes by flooding a task in the network. While the task is being flooded, sensor nodes record information on the nodes which forwarded the task to them as their gradient. Alternative paths from sensor nodes to the sink are established. This approach is referred as direct diffusion.

The sensor node processes a query upon its arrival. If the node can resolve it the results are disseminated. Proposed middleware systems for processing user queries include TinyDB, COUGAR and ACQUIRE.

The current research span efforts in distributed data collection with centralised storage system for data persistence. To achieve the full vision of seamless cooperating objects a shift is required from this scenario to the distributed data collection and storage, where some powerful cooperating objects will be responsible for cooperative data storage. Support for mobility and heterogeneous systems are also important aspects to be considered.

In particular, if data can be cached or deposited at specific nodes in the network (i.e. in infrastructure nodes), the overall system reliability may be significantly enhanced. However, this scenario implies an ability to place responsibility (e.g. for buffering) at specific locations in the network.

7.1.10 Motion Planning

Description and Relevance

Motion planning involves determining the motions of mobile nodes so that they reach a goal state by optimising a criterion, such as the minimisation of the travelled distance, and without colliding into obstacles. This research area has been widely studied in robotics in the last decades and many successful algorithms and techniques have been developed and used in many applications. Furthermore, reactive techniques for autonomous navigation toward a goal avoiding obstacles have been also widely studied and applied in robotics.

In the framework of Wireless Sensor Networks with mobile nodes, motion planning is an important issue that has not been sufficiently addressed yet. Existing motion planning algorithms require an environment model, i.e. a map, and a kinematic/dynamic model of the mobile node. If the environment model is not available or it can change, it is necessary to apply map-building techniques which require suitable sensors. Furthermore, the computational complexity of the algorithms for motion planning and map-building are usually beyond the capabilities of the hardware on-board the mobile nodes of a Wireless Sensor Network. Therefore, those algorithms should be customised to be executed properly on the mobile nodes.

Applications of motion planning could include the optimal use of mobile nodes as "data mules". Then, the problem is the computation, possibly in a distributed way, of the paths to be followed by the mobile nodes to visit the location of static nodes, or cluster of static nodes. Updating of this path depending on changes of the environment or using new data collected by the sensor network should be also considered.

Moreover, communication constraints are an essential issue in the design of the motion planning algorithms for the mobile nodes of a Wireless Sensor Network, requiring novel approaches that are not considered in classical path planning methods in robotics.

Existing Trends

Recently, a new research field which considers the information provided by a static Wireless Sensor Networks to guide a mobile robot (ground and aerial) in a given environment has raised. It is an emergent trend in the robotics community and very few algorithms have been developed and tested up to now. The idea behind this trend is to consider the sensor nodes deployed in the environment as an extension of the sensorial capabilities on board the mobile robot. An extreme case that has been currently considered is the guidance of a robot with no sensors on board (just using the information from a Wireless Sensor Network).

This trend could be considered as a starting point towards the more ambitious objective of developing motion planning methods for the mobile nodes of a Wireless Sensor Network. Those methods should take into account the communication constraints of the nodes and the relation with self-hierarchical organisation and/or clustering techniques. To the best of our knowledge, this second trend has not started yet in the Wireless Sensor Network or robotics communities, but seems an important requirement to improve the performance of a Wireless Sensor Network and should be addressed in the next future.

7.1.11 Multiple Sinks

Description and Relevance

Most algorithms developed for Cooperating Objects assume only one sink or base station. If more than one sink is used, these algorithms do not work properly any more. However, depending on the application scenario, certain functionality is required to allow for the use of several sinks at the same time.

Routing can benefit from a hybrid network with externally connected sinks. To send packets from the network to an external user usually the nearest base station is best. Vice versa, to send a packet to the network the base station nearest to the destination should be selected. If a packet is forwarded within the network the packet can be routed completely in the network or part of the routing can be done via two base stations and their external connection.

The implicit clustering of the network of cooperating objects could be reflected in their addressing, but for this new algorithms and mechanisms that take into account the costs of going "outside" of the Wireless Sensor Network need to be developed.

Other algorithms have to be examined if they are affected by a routing protocol that tunnels packets via an external network. In the terms used by the security community, a "wormhole attack" is very similar to this kind of – in this case desired – routing behaviour. Therefore, the design of the protocols and their interactions have to studied carefully.

In Cooperating Objects that are large or tend to be fragmented, the use of multiple sinks is beneficial. For example, packets can be routed via the external network, the sinks are connected with, near the target area which is usually much faster, saves energy and implies a lower loss probability. Such cooperating objects are used in many scenarios, e.g. in logistics, transportation, and their spread is likely to increase.

Existing Trends

Several routing protocols as well as some data-centric protocols such as TinyDB have been proposed so far that are able to cope with the use of multiple sinks. In terms of routing, they are able to divide their streams into several coordinated data streams. In some cases, even mobile sinks are supported. In the case of data-centric protocols that normally deal with queries injected in the Cooperating Object network in search of a solution, the use of multiple sinks implies the need for optimisation techniques that benefit from other independent data streams that might be computed through the same path.

Although some research has started in this field already, most researchers share the opinion that there is enough potential with the use of only one sink and that multiple sinks should be dealt with at a later point in time.

7.1.12 Scalability

Description and Relevance

The number of nodes in a network may vary depending on the need of the application. Scalability is the efficiency of a system performance under varying number of nodes in a network. The consistency and stability of the system can only be maintained by using scalable algorithms. In this manner, researchers often try to provide their solutions to successfully and efficiently work with a great number of nodes by developing scalable algorithms.

Since the applications and scenarios, considered in wireless sensor networks context, are generally built using multi-hop protocols, the number of nodes is expected to be in order of hundreds or even thousands. This is especially the case in outdoor scenarios. In such networks, algorithms must be created to work in large scales. The difficulty in running and maintaining the consistency of the whole sensor network brings up the necessity of a great effort to create scalable solutions. When speaking about Cooperating Objects, those sensor networks may not only be set up with a huge number of nodes, but many of new nodes can join the system. Installing new adaptive software on each node is very difficult and costly. Hence, the software, which is initially used, must be adaptive to be able to deal with varying loads and the updates must only be done only when the duty needs to be changed. Even if there is no nodes expected to join to the network after the initial deployment, scalability is essential when thinking the total lifetime of the system. Increasing the lifetime is one of the main issues in wireless sensor networks. With efficient scalable approaches, total lifetime can be increased by decreasing the total energy spent.

Existing Trends

From the very beginning of Wireless Sensor Network and Cooperating Object research, people has been aware of the importance of scalability in the network. The algorithms are developed to work at a large scale, as much as possible. However, existing approaches are still far away from the optimal desired scalability. One way of fulfilling this need is temporarily eliminating some nodes from actively joining the data propagation. In many proposals, the network is divided into subnetworks and those subnetworks are attached to each other. Instead of building a topology of hundreds of nodes forming a single graph, the subnetworks build their own structure and send the concatenated data as if each subnetwork was a single node in the whole topology. Another trend is putting some nodes into sleep mode and decreasing the total number of nodes working at a time. Covering the operation area with the minimum number of nodes is very important, but achieving this is very difficult, especially if the deployment is random. Putting some nodes into sleep mode makes the network behave as if it covered the whole area with the optimal number of nodes.

7.1.13 Quality of Service

Description and Relevance

Quality of Service (QoS) has been (and still is) a major research topic in wired and wireless telecommunication networks. In such traditional contexts, QoS-support is generally intended as the ability of the communication system to guarantee some specific performance parameters, such as end-to-end packet delay, delay jitter, and packet loss rate. The most prominent and important topic *Real-time* is discussed in section 7.1.18.

The main difference to traditional wired networking is that in such systems the network administrator controls the channel because he owns the wire, but wireless Cooperating Objects operate mostly in a licence-free wireless band. Thus, the channel really is shared and QoS "guarantees" have to be different. Therefore, Cooperating Object systems extend the concept of QoS by requiring that transmitted data is *properly* received by the destination, where the actual meaning of *properly* depends on the specific application requirement. More specifically, the frontier of investigation is extended in two directions.

On the one hand, the functionalities that have to be developed to realize the cooperation among objects involve novel requirements which, in turn, enlarge the concept itself of QoS. For instance, queries issued to a cooperating-objects system might include important Vertical Function parameters. These are application-dependent and even query dependent parameters: (a) for a query to be successfully resolved, the sensors must deliver fresh data, (b) the query must be answered fast enough, and (c) the precision with which the query is answered must meet the query requirements. This can be regarded as new concept of quality of service requirements that remains to be explored.

On the other hand, the provision of QoS services over networks composed by heterogeneous objects rises a number of challenges that need to be addressed in the next future. For instance, the support of QoS over Wireless Sensor Networks or vehicular networks will require the definition of new mechanisms capable of coping with the peculiar aspects that characterise the underlying enabling technologies.

Existing Trends

The main trend in this context concerns the evolution of the QoS paradigm, which is now being extended to consider, besides classical performance parameters, other specifications associated to the novel application scenarios enabled by Cooperating Object-systems. For instance, QoS in Wireless Sensor Networks for environmental monitoring is now being intended in terms of event notification rate at the sink node, rather than successful packet delivery rate. Some of the parameters that are being included in such an extended framework comprehend missing event-detection probability, energy efficiency, data accuracy and many others.

The research community is aware of this evolution on the concept of QoS-provisioning. In response, solutions to enable QoS support over heterogeneous platforms are being investigated. Once again, the general trend considers a cross-layered approach for the design of the algorithms. Breaking the layering paradigm, permits to control the interaction among mechanisms operating at different layers, thus improving the effectiveness of the solutions in supporting the required QoS performance. Such an improvement implies a trade-off in terms of flexibility. Indeed, the software structure becomes more difficult to update and maintain.

There is, therefore, a trade-off between flexibility of the protocol stack and efficiency. Nevertheless, the performance gain that can be potentially achieved with a cross layer approach, generally overcomes the drawbacks of a more complex protocol structure, specifically in the context of cooperating object systems. Indeed, as also stated in the conclusions of section 4.2, Cooperating Objects may present irreconcilable discrepancies, which require customised solutions. Therefore, the layered structure, which would facilitate the interconnection of heterogeneous systems, might be sacrificed for a greater flexibility regarding the design of the protocol stack.

7.1.14 Robustness

Description and Relevance

In a Cooperating Object application, the operational and environmental conditions may not be especially favourable. Generally speaking, devices in such application domains such as sensors, actuators, etc., should be resistant to the potentially harsh environmental conditions such as high and/or low temperature, other RF devices, etc., and should integrate seamlessly in the environment. On the other hand, robustness is not only an issue for node hardware but also for algorithms running on the network. Various software algorithms used for routing, localisation, etc., should keep working properly even if operational conditions or the structure of the network changes. For both hardware and software design steps, robustness must be taken into consideration.

Some applications of Cooperating Objects such as environmental monitoring, security and surveillance, and disaster relief may be deployed in hostile environments and needed to last for years. It is very challenging to encapsulate sensors from the real world particulary in aqueous environments. Also, once the nodes are disseminated, it may be impossible to attain the network again. In such applications, nodes must be designed to be resistant against harmful effects of the environment and having as long as possible life time.

On the other hand, materials used for robust sensor design may harm the flora, fauna or the ecological structure of the environment. Therefore, the designers must take this into account. Also algorithms used throughout the network must be resistant and adaptive to sudden and/or long-term changes, that is, apart from the hardware, the software of the wireless sensor networks must be robust.

Existing Trends

There are quite a few research approaches for the design of robust software and hardware for wireless sensor networks, and in recent years, significant advances have been attained. For some indoor and outdoor applications where the number of sensors is limited and the environmental conditions are not tough, off-the-shelf sensors providing robustness with good housing methods and software for fulfilling the required task are available. On the other hand, as the application requires more sensors and the operational environment gets more hostile and wild, the possibility of realizing the application with the available hardware and software decreases. Considering the software of the network, tested and approved robust multihop algorithms such as routing, localisation, etc., are not available yet. Also sensors robust to each kind of environments and regarding the ecology have not produced yet.

7.1.15 Mobility

Description and relevance

Mobility is an important characteristic identified in some of the application scenarios discussed in previous chapters and in the Embedded WiSeNts studies. In sentient applications, Cooperating Objects can be moving relative to each other and to the observer. In other cases, the mobile observer device attached to people, animals or vehicles travels over an area periodically in order to collect, store or forward sensor data (Mules model). Traffic monitoring applications are examples of such a model, where vehicles sense the environment and forward sensor data to nearby vehicles or deployed base stations. Also, the phenomenon itself can be moving relative to the cooperating objects and observer. This mobility pattern can be found in detecting animals by sensing some their characteristics (e.g. sound of birds).

Mobility models are often described with respect to the patterns followed by mobile nodes, their speed and movement direction, presence or not of obstacles and the radio propagation models used. Scenarios may fall within two broad types of mobility.

High mobility: Refers to the motion of nodes at high relative velocities. An example is the inter-vehicular networks used to address critical road safety and efficiency. For instance, coordinated collision avoidance could significantly reduce road accidents.

Despite the constraints on the movement of vehicles (i.e., they must stay on the lanes), the network will tend to experience very rapid changes in topology. In some of the vehicle network applications communication happens with destinations that are groups of vehicles. Thus, the traffic pattern is predominantly multicast, in particular geocasting with physical areas of coverage, while the data rate is generally rather high (car-to-car voice/video connections, web surfing, and so on).

Low mobility: Concerns the communication scenarios between people through peer-to-peer communication or through an Internet gateway in 802.11 wireless networks. The mobiles nodes are PDAs, phones and laptops with reasonable resources available. Low-resource wireless nodes can be used as well. Personal area networks (PAN) and home networks based on Bluetooth, ZigBee and UWB radio systems are designed for low mobility applications.

Mobile ad-hoc networks (MANETs) have been examined in the literature to support the requirements of low and high mobility applications. Examples of algorithms, paradigms and architectures were explored in chapters 4.2 and 4.4. The mobility model itself is an important part to understand experimentally (simulation or testbed experiments) the issues of the applications.

Although mobility depends on the application, certain properties define a mobility scenario. Speed, obstacles and radio propagation models are important aspects. Other key issues include scale of nodes, network density and network partitioning. These factors give rise to some research gaps:

Comparison of architectures: Hybrid network architectures similar to that of 3G or mesh networks (wired and wireless infrastructure)ireless/wired infrastructure versus purely ad-hoc systems with no pre-planned infrastructure nodes. Many proposed MANET applications appear to be communication among and between nodes where no pre-deployed infrastructure is available. This includes military (battlefield) units, emergency response scenarios, and exploration in dangerous or unknown areas. Multihop MANETs seem only acceptable with small number of hops otherwise the performance is extremely poor. Many other applications that have been proposed appear to be supportable at reasonable cost with various types of pre-planned infrastructure (e.g. 3G network).

Providing quality of service to user applications. This is an issue particularly in scenarios with unreliable wireless links and user/node mobility. QoS refers here to the application tolerance to latency, jitter, unreliability and bandwidth variation. This problem is further complicated when no pre-planned infrastructure exist.

The mobility of some of the devices have been explored by architectures such as the data mules, in which a group of mobile nodes move in the deployment area collecting data from the sensor nodes and delivering the collected data to the sinks. In this case, there is no guarantee on timely data delivery. In contrast, critical applications such as patient monitoring require strict bounds on latency and guaranteed data delivery.

Need for a set of benchmark datasets and more realistic mobility models. Many available results are not comparable, since no common set of measurements data exist. To enable better research, mobility models and benchmarks should be used to evaluate communication protocols and middleware approaches. A framework is required to represent the benchmark datasets so that they can be shared, e.g. for system evaluation and testing. This representation should comprise mobility models derived from real-world data with a combination of some of the following characteristics: user/node mobility, traffic characteristics, network topology, link quality, distribution of nodes – to name a few.

Support for geographical routing. Location-based routing with geographical coordinates and mobility management has been identified as a potential solution to the issue of communicating data among mobile cooperating objects. Such approaches, however, assume that a location service is in place to keep and inform the position of a given node. There are a few proposed location services but none of them provide a scalable and distributed service.

Also, positioning is still inaccurate, better algorithms are needed. Satellite-based positioning system (e.g. GPS) are capable of offering a quasi real-time positioning service.

In contrast, location algorithms that are based on transmitting beacon signals are not suitable for strictly real-time applications - often they require the reception of several control packets to reduce the estimation error.

Security: node anonymity and privacy regarding how nodes move in space and time may be required.

Coordination among mobile Cooperating Objects. The use of a team of mobile nodes require coordination for optimal coverage of an area, for instance, to avoid collisions in road safety applications.

Motion planning. An important trend is related to the study of how a sensor network can compute, in a distributed way, the path that a mobile cooperating object should follow. This path can be updated depending on the changes of the environment (e.g. mobility of observers, other Cooperating Objects or the phenomenon). More algorithms and theoretical studies are needed in this area.

Existing Trends

The routing protocols for Wireless Sensor Networks are generally designed for networks that have fixed homogeneous sensor nodes and are based on the assumption that all nodes try to convey data to a central node or one of several backbone nodes. However, in cooperating objects networks there will be heterogeneous nodes that can be mobile, and the sensed data will be needed by many nodes, i.e., multiple sinks.

Generally speaking, the majority of these algorithms can copo (although not efficiently) with changes of the topology due to node mobility. Most of them, however, react to topology changes by dropping the broken paths and computing new ones, thus incurring a performance degradation.

In particular, mobility may affect cluster-based algorithms because of the cost for maintaining the cluster architecture in a set of mobile nodes. Routing algorithms specifically designed for networks with slow mobile nodes are, for example, GAF and TTDD, which attempt to estimate the nodes trajectories. Other suitable protocols for low mobility are the SPIN family of protocols since their forwarding decisions use local neighbourhood information.

Mobility might represent an issue for MAC protocols as well. First, mobility involves topology variations that may affect algorithms that need to tune some parameters with respect to the density of nodes in the contention area (SIFT, TRAMA, TSMA, MACAW). Second, MAC algorithms based on medium reservation mechanisms (MACA, MACAW) may fail in case of mobility, since the reservation procedures usually assume static nodes. For instance, algorithms based on the RTS/CTS handshake to reserve the medium may fail because the nodes can move outside the mutual coverage range after the handshake or external nodes can get into the contention area and start transmitting data unaware of any medium reservation.

7.1.16 Security

Description and Relevance

Currently the use of Cooperating Objects is gaining momentum and we envision it to become a commercial success within the next few years. Since they will be included in business processes, Cooperating Objects have to be secured in several ways to ensure their reliable and trustworthy operation. Therefore, security is one of the key points for the acceptance of Cooperating Object technology outside the research community.

Computer security traditionally tries to create a computing platform where users or programs can only perform operations they are allowed to. For Cooperating Objects, this goal cannot be achieved for a single node only since they are often physically accessible to attackers. They could read even the entire contents of main memory or install their programs without relying on functions of the operating system. The focus of security measures in this area is to secure the Cooperating Object as a whole while tolerating the attack of single nodes. The use of tamper resistant devices is no option because such devices are too expensive.

If a single node cannot be protected, the data sent between the Cooperating Objects should definitely be kept private. On the one hand, data has to be authenticated, i.e. an object can assure that the data was sent by another specific object. On the other hand, data has to be encrypted in order to prevent unauthorised listening. Both can be accomplished using pairwise symmetric keys or a public-key infrastructure. Public-key approaches based on prime numbers require too much computation to be efficiently executed on small devices. For symmetric keys, the main problem is the distribution or the establishment of pairwise keys.

In some cases, encryption is not enough to ensure secrecy since individual data packets might reveal some piece of information that, when put together with other packets, could reveal enough information to make the system insecure. Therefore, it is crucial that important messages be combined with dummy messages. The disadvantages of this method are higher power consumption and an additional source of latency for important messages.

Another security aspect is the question if the data delivered by a Cooperating Object is valid since an attacker can take over a node including its keys and send wrong but authenticated data. Detection methods include random sampling mechanisms which can also be applied to aggregated data. Detected misbehaviour must be evaluated in cooperation with other nodes using voting or reputation systems.

Careful protocol design is also needed since encryption is not sufficient. Usually, researchers develop protocols without security as a requirement but think that security can be added later on. As has been show, this cannot be done for all protocols or only with a big effort.

Existing Trends

Concerning encryption, elliptic curves are investigated as a promising approach for small Cooperating Objects, but are far from being a standard. Symmetric cryptography requires efficient methods for the distribution of keys and, although a good amount of effort is put into this research area and promising results are starting to become available, there is a need for more research in this area. Data integrity assessment, voting systems, and reputation frameworks for Cooperating Objects have received more attention in the last years, but are still in their infancy. Finally, some routing protocols especially designed with security in mind exist in the literature, but they are not normally deployed in current projects and are seldom used in research.

7.1.17 Heterogeneity

Description and Relevance

The integration of heterogeneous objects with different embedded information processing and communication capabilities has a huge number of application possibilities. Furthermore, Cooperating Objects with heterogeneous hardware offer the additional advantage of exploiting the complementarities and specialisation of each object. For instance, the integration of Radio-frequency Identifier (RFID) technology with Wireless Sensor Networks provides a symbiotic solution that leads to improved performance of the system. Also, the cooperation among heterogeneous autonomous objects, such as robots and Wireless Sensor Networks, or Wireless Sensor Networks and inter-vehicle networks, as well as the integration of such systems with other communication systems (e.g. cellular), may result in a dramatic increase of the domains of applications that might be provided to the users.

Most objects with embedded processing and communication capabilities have been designed ad-hoc to optimise particular functionalities taking into account technology constraints existing at design time. The results are objects with heterogeneous hardware and fixed modes of programming and operation. Then, their integration to operate as a team has significant limitations. Furthermore, heterogeneity increases their complexity and, in fact, most techniques for the analysis and design consider homogeneous teams of objects with the same information processing and communication capabilities.

Another source of heterogeneity is mobility. For example, it has been pointed out that the consideration of the interactions between static and mobile objects in a Wireless Sensor Network is also very interesting for applications. However, the consideration of mobile nodes increases again its complexity.

Furthermore, efficient cooperation of autonomous systems requires the consideration of perception, planning and control activities in the framework of a theory of operation in which both local and global approaches could be appropriately integrated. Existing techniques mainly consider homogeneous teams and representations, but fails when considering heterogeneous systems.

Today there is a large variety of objects with embedded capabilities designed to perform different functions, including sensor nodes, mobile phones, PDAs and complex machines with significant processing and communication capabilities. The integration of these objects offers unprecedented possibilities of application in homes, civil security, traffic management and industries, among others.

Thus, for example, the integration of mobile devices such as PDAs and mobile phones in Wireless Sensor Networks opens a large number of new applications. Moreover, the analysis of the state of the art has pointed out that the use of mobile nodes in a Wireless Sensor Network offers significant advantages such as less number of nodes to cover the same area, dynamic adaptation to the environment triggers or changes, and dynamic change of the topology to optimise communications in the network.

Autonomous systems have been applied to perform activities where the human access is costly, difficult, dangerous or even impossible due to physical limitations (i.e. micro and nano scales). These systems can extend the human capabilities to perceive and actuate in inaccessible places, remote locations or polluted environments. There are many tasks that require the cooperation of multiple autonomous systems because their complexity (i.e. multiple actions are required at the same time in different locations) or because uncertainties (i.e. redundant information is required to improve reliability). Moreover, the interest of the integration of these autonomous systems with Wireless Sensor Networks has been also pointed out in chapter 4.

Existing Trends

The use of gateways to communicate between different types of entities has been widely used in existing networks of heterogeneous objects. However, the exploitation of the possibilities of the Cooperating Object technology requires better solutions. Thus, the development of the middleware making the communication between heterogeneous objects fully transparent to the application designer has been recognised as a major trend to enable the development of systems of networked objects. Furthermore, it has been pointed out that there is a need of new generic real-time architectures that consider the cooperation of heterogeneous objects at several levels.

The middleware will be the baseline to develop new network-based functionalities for the cooperation between objects with an appropriated balance of flexibility and individual performance of the members of the team. These functionalities include network-based cooperative perception and cooperative control strategies with a large number of potential applications.

A key issue is the development of new models and cooperation paradigms to deal with the complexity involved in these systems. Therefore, in spite of the techniques that have been developed in the last 10 years for low level sensor data fusion, the integration of higher perception levels still requires significant efforts. In fact, most cooperative perception techniques are based on probabilistic approaches to integrate the information provided by a relative low number of homogeneous objects. The integration of a large number of heterogeneous sensors and individual perceptions poses a number of theoretical and practical problems. Thus, for example, there are not enough methods and technologies to enable the appropriated integration of wireless sensor nodes with camera-based systems to perceive the environment, particularly when considering fast dynamic evolving environments. Furthermore, the mobility of the sensor nodes and cameras pose additional changes.

A further step is the consideration of sensor-actuator networks involving the generation of actions. Many distributed control problems involving heterogeneous entities are unsolved, particularly when considering tasks involving coordination at a global and a local level. Traditional solutions are typically centralised. If one attempts to control these systems in real-time using standard control design techniques, severe limitations will quickly be encountered as most optimal control techniques cannot handle systems with a high numbers of dimensions and with a large number of inputs and outputs. It is also not feasible to control these systems with centralised schemes, as these require high levels of connectivity, impose a substantial computational burden, and are typically more sensitive to failures and modeling errors than decentralised schemes. In the last years distributed control systems have been also proposed. However, most existing distributed control techniques still consider teams homogeneous objects as a basic modelling assumption.

The same happens when considering task planning in teams of cooperating objects. The analysis of the state of the art has revealed that there is also a lack of formal methods to address the optimal cooperation of heterogeneous objects exploiting the complementarities of these objects. In general, it can be seen that there is a lack of theories that allow a holistic analysis of the heterogeneous team (individual members, interactions, subteams and overall system), involving all the perception and action capabilities of each object.

7.1.18 Real-time

Description and Relevance

An application, or a process, is said to be a real-time application when it requires a predictable temporal behaviour: more precisely this means that it requires explicit tasks to be completed before a given deadline. For example, a real-time constraint in a network might be: "this event must reach a sink node 100ms (at the latest) after having been raised". Usually, two classes are distinguished, namely *h*ard real-time applications and *s*oft real-time applications.

Hard (or strict) real-time means that missing a deadline leads to a critical or catastrophic failure in the application domain; hence, temporal constraints must be strictly respected to ensure the reliable operation of the application. An example of a strict real-time application is the ABS car breaking system. Soft real-time means that the application can survive or tolerate missing some deadlines; a typical example would be multimedia streaming over a network. A soft real-time system tries to minimise the miss ratio, or to provide a QoS guarantee on the miss ratio.

The general principle of real-time systems design is to ensure temporal predictability of the tasks involved in the application, and in their scheduling. Hard real-time systems require a strict worst-case execution time (WCET) analysis of the tasks (and the related worst case transmission times for the communication aspect), while soft real-time systems can use statistical analysis based on code profiling.

A fundamental difficulty in designing systems that would offer real time guarantees in the context of cooperating objects is that each of these classes use antagonist design principles:

On the one hand, real-time systems use over-allocation of resources (such as in pessimistic WCET analysis), and usually prevent dynamic behaviour (and at first allocation of resources) to ensure predictability.

On the other hand, Cooperating Object systems, which rely on autonomous and resource constrained devices, try to optimise resource usage, and also depend heavily (by definition) on dynamic set-ups due to the random nature of deployment, and thus, dynamic behaviour. An example is setting up a virtual resource from a set of sensors, such as computing the average temperature in an area, in order to meet a given accuracy requirement.

Real-time constraints have not been studied much in the context of cooperating objects. However, they are especially important for many sensing applications: real world processes and phenomena often require real-time data acquisition and processing. Some classes include mission critical applications, such as early warning systems for natural disasters or contamination (forest fires, earthquakes, tsunamis, radiation, etc.) or support for emergency interventions (firemen, etc.) for which real-time guarantees are mandatory. These real-time constraints are even more strict in applications such as machine fault-detection and control, vehicle control and transportation systems.

Existing Trends

Real-time properties are needed both at the node level (hardware and software), but also at the network level. The former can be considered as designing real-time embedded systems with harder hypothesis due to resource limitations (energy, computing power and limited memory). The important open question that research has to investigate is to determine whether the classical approaches of embedded real-time systems (such as formal WCET analysis, synchronous languages) can be applied to cooperating objects such as wireless sensors despite their strong resource limitations.

The latter raises even more challenging issues: the communication aspect in Cooperating Objects involves a lot of dynamic behaviour and thus unpredictability. Such systems would need both real-time MAC layers and real-time multi-hop routing protocols. Although some soft real-time MAC layers have been proposed (802.11 DCF, RAP, SPEED), to the best of our knowledge there is only one hard real-time MAC protocol: I-EDF (implicit earliest deadline first). Unfortunately, this protocol assumes a static cell-based deployment of the nodes, which is not compatible with many of the scenarios described in this document of Cooperating Objects.

Finally, regarding dynamic multi-hop communications, no solution provides the level of guarantees expected for hard real-time systems and there is no formal validation of such approaches in the context of Cooperating Objects.

7.1.19 Adaptive Systems

Description and Relevance

Adaptive systems can be divided into two categories: First, the functionality of the application itself is adaptable, e.g. a mobile phone might change the way incoming calls are handled if the user is in a meeting. Secondly, the functionality remains the same from the point of view of the application, but some underlying algorithms change to behave optimally in a the current environment.

Especially the second class of adaptation is important for Cooperating Objects, since the requirements and the environment of applications are dynamic and change often. Most algorithms used in cooperating objects are suitable for a subset of requirements and conditions only. Therefore, the developer of a system has to provide many algorithms and decide when to use which one. This cumbersome task can be taken over by adaptive systems.

Such systems have to support different adaptation models since the goals of the adaptation may also change over time. If more than one model is available to the adaptation system the best suited can be selected automatically. No system supporting different models has been developed so far.

Tightly coupled with adaptive systems is the cross-layer design of algorithms and the system support for such a design. Cross-layer interactions are used widely for optimisation purposes. Sometimes, they are necessary to deal with special properties of sensor networks. If an adaptive system exchanges algorithms it has ensure that cross-layer data which is needed by other algorithms is still available. Additionally, it has to provide means to store this data outside of the algorithms so that the data is not lost when the algorithms changes.

Adaptive systems are also used to assign resources to the tasks of a Cooperating Object. This resource management as well as the adaptation of algorithms requires multidimensional optimisation techniques. Furthermore, if mobile nodes are considered, this includes the application of motion planning techniques (see previous gap on motion planning). Their usage has not been examined well yet in the field of Cooperating Objects.

Existing Trends

First, adaptation systems were designed for more powerful cooperating objects. Now, adaptation frameworks for low-resource devices are emerging. In these frameworks, the system-driven adaptation of applications to the context is investigated, i.e. systems that manage the adaptation transparently. Support for cross-layer interactions are also a part of such systems and include extensions to programming languages and operating systems as needed for the seamless integration of cross-layer optimisation techniques.

7.1.20 Operating Systems

Description and Relevance

Cooperating Objects comprise a variety of devices, from special-purpose sensor nodes to general-purpose PDAs. The operating systems used for Cooperating Objects have, therefore, a similar spectrum. TinyOS, for example, is designed for the constraints of sensor nodes. The main purpose of such operating systems for Cooperating Objects is to provide an abstract interface to the underlying hardware and to schedule system resources. Operating systems for hand-held devices offer more capabilities, e.g. memory management or a graphical user interface. Since several different devices are expected to work together in Cooperating Objects which requires partly the same functionality questions of interoperability and portability arise. Currently, no minimal standard for operating systems in the Cooperating Objects domain exists applications can be built upon.

Especially in the *Control and Automation* domain, real-time requirements have to be met. Many real-time operating systems exist but they are used hardly in cooperating systems context. FreeRTOS, for example, is available for many microcontrollers in Cooperating Objects, but not tailored to the specific Cooperating Object platform. Most of the commonly used sensor network operating systems do not provide real-time guarantees. This makes the development of real-time applications even harder since many reusable modules are coded for other operating systems only.

Operating systems are traditionally service centric, i.e. they offer clear interfaces with functions the application can call. Cooperating Objects are often regarded as data centric. Thus, there is a gap between the application functionality and the operating system. The developer has to think both data centric and service centric to bridge it by translating the data centric processing into service centric calls. Data centric operating system can shorten this gap and assist the developer.

Existing Trends

Real-time operating systems with a few hundred bytes of ROM consumption have been developed. We expect them to be used for sensor networks soon. An operating systems with a data centric architecture has been developed as well but experience with this kind of operating systems is missing. We expect this to be an interesting field of research in the future.

Although layered architectures facilitate key design properties such as flexibility and abstraction levels, they require attention to conformance and can severely impact performance in systems that lack hardware resources. To achieve an energy-efficient design, the traditional strict modularisation or layering is not appropriate. An approach that can be used to improve performance in systems low in resources is to implement cross-layer interactions where the software components do not necessarily interact with components immediately above or below in the abstraction level. First approaches to ease the development of cross-layer software with support from the operating system have appeared.

7.1.21 Programming Models

Description and Relevance

According to the ON World Inc. report on Wireless Sensor Networks dated Nov 2004 [50], one of the major barriers to adoption of sensor networks is the lack of ease of use. Indeed, implementing and deploying a sensor network today is a tedious task whose successful completion typically requires significant experience and expertise with the underlying sensor node hardware and operating system. This is in stark contrast to the anticipated use of large-scale sensor networks for real-world applications, where application-domain experts – rather than system experts – should be able to customise and deploy sensor networks.

One of reasons for the above problem is the lack of proper programming models. Currently, most applications are implemented on top of simple operating systems such as TinyOS. However, such software provides a very system-centric view, where the developer has to deal with low-level issues such as simplistic memory and concurrency models. Moreover, sensor networks constitute very dynamic distributed systems, where the developer has to take care of node and communications failures.

Novel high-level programming models are needed to raise the level of abstraction from system-centric programming to application-centric programming of sensor networks. Such programming models should enable application-domain experts to write applications by specifying the expected behaviour rather than by having to deal with system details. In particular, such models can support programming a sensor network as a whole (rather than individual nodes) using declarative specifications of the functionality that needs to be realized.

In order to improve the programming process further, graphical user interfaces are necessary that support the programming model. For example, simple decision trees could be drawn or queries could be build. The network of Cooperating Objects could be pre-planned and task could be assigned to specific nodes. Finally, the resulting network and the program can also be checked against the model to find possible errors. Section 7.1.24 deals with installing, testing and debugging.

Existing Trends

Examples of such high-level programming models can be found in the literature. For example, TinyDB enables programming sensor networks by specifying a declarative SQL query over the whole network. Generic Role Assignment allows the specification of declarative rules to achieve a network-wide assignment of functions ("roles") to sensor nodes. However, these examples cover only very specific problem domains rather than proving a generic solution. We not only need high-level programming models for a larger set of problem domains, but also mechanisms for gluing together a complete application using multiple programming models (e.g., to pose a query only over nodes that have been assigned a particular role).

7.1.22 System Integration

Description and Relevance

In testbeds or experimental deployments, Cooperating Objects are regarded separately, i.e. the developer interfaces directly on a low abstraction layer with them. However in an operational deployment, the Cooperating Object has to be included in a bigger context of – mostly existing – frontend software that can, for example, control or query the Cooperating Object and receives in turn notifications or answers to the queries. The communication between the Cooperating Object and the frontend software involves several heterogeneous intermediate systems; for the heterogeneity of the Cooperating Object itself and the possible distribution of task inside it refer to the *Heterogeneity* gap (Section 7.1.17).

The problem becomes easy to perceive when it is analysed from different perspectives. An example of a system integration issue occurs in medical applications. First, sensors (e.g physiological and blood sampling) need to be integrated. Suitable digital and analog interfaces and the software that allows their integration need to be in place. The second step is the integration of sensor nodes in a personal area network, where sensors are likely to be attached to different parts of the patient's body. Data collected from such sensors are forwarded to a base station where the data has to be integrated into existing medical infrastructure.

The routing of messages between the Cooperating Object and the frontend software is usually done via existing infrastructure, for example Ethernet or Wireless LAN. In contrast to the data-centric nature of Cooperating Objects, such networks use address-centric routing. The gateways between both "worlds" have to take care for the correct translation of the addressing modes. More Cooperating Object logic can be put to the gateways if the Cooperating Object wants to make use of the infrastructure, for example to send messages to other parts of the same Cooperating Object in a cheap way.

Regarding the data interface between the frontend software and the Cooperating Object more conversion has to be done. Small Cooperating Objects normally use proprietary but space saving data structures to limit the amount of data to be transmitted using wireless technology. Most frontend software can read and write data in XML format. Thus, the commands and the data have to be converted between both formats.

To improve the performance of the whole system, program logic usually found in the application can be pushed towards the Cooperating Object system. With this approach, for example the frontend does not have to evaluate all the data, but the Cooperating Object selects the relevant information to be sent to the frontend.

Existing Trends

Real applications will probably require advanced sensors to measure complex physical phenomenon (e.g. water pipes and medical devices). It is likely that the emphasis will be put on the design of more complex micro-electromechanical (MEMS) sensors. To support this development, a cooperating object system will be built from powerful resources that need more implemented functionality and efficient interactions than the ones currently offered by wireless sensor operating systems. It is also unclear today how they should be integrated into a distributed sensor system. The choice of interfaces for the integration process is rather important. For industrial sensors, there are ongoing standardisation actions, e.g. IEEE 1451.

The current trend in embedded systems is to adopt a component-based approach. The component 'wiring' process will facilitate the synthesis of individual components into a larger-scale system. The Cooperating Object system may be arranged in several levels of abstraction from the most abstract (closest to the application) down to the most concrete (closest to the hardware devices). Abstraction allows better software structuring for clarity and reuse. A framework that supports the smooth integration of independently developed components is needed.

Several real-world Cooperating Objects have been deployed and connected to frontend systems recently using a multi-tier architecture, but each of them was designed separately. We expect a general software framework to appear, with standardised interfaces for the integration of general and application-specific components. Such an architectural design should enable the rapid deployment of new services. From the system developers point of view, the development cycle of Cooperating Objects should be reduced with improved design methods and tools that support incremental or novel functionality. For example, for traditional building automation, the LonWorks platform was established which provides among other things a communication protocol between the single components and standardised interfaces using Web Services.

7.1.23 Modeling and Analysis

Description and Relevance

Most of the algorithms proposed in the literature in the context of cooperating objects systems have been evaluated by means of theoretical analysis or computer simulations. In both cases, the analysis is usually carried out under very simplistic hypotheses, i.e., referring to simplistic models for radio propagation, nodes architecture and device capabilities. As a result, several solutions proposed in literature hardly work when deployed in the real world, as it has been observed in the few empirical-oriented works that have been recently carried out in real test beds. Consequently, in the next future the research in Wireless Sensor Network should be more experimental-oriented, in order to verify the limits of the theoretical analysis and to reveal the issues that inevitably arise when a system is really deployed in the field. This knowledge can, hence, be used to refine existing solutions and develop novel approaches. In parallel, more realistic channel, environment and object models ought to be designed, in order to permit more accurate algorithm design and testing.

Among the aspects that need more accurate modeling, we recall the energy cost of data processing. In fact, several algorithms make use of some data processing techniques to perform data fusion, data aggregation, routing and so on, in order to maximise the efficiency of the network and reduce the power consumption due to signal transmission/reception. However, the computational and energy cost of these processing is rarely considered, despite it might have a relevant impact on the overall energy budget of the system.

Similarly, the literature usually considers very basic and oversimplified models for the node architecture. For instance, in Wireless Sensor Network works, nodes are very often described as *limited in energy, storage capacity and computational power,* but the actual value of such elements is rarely specified. The result is that most of the solutions provided in the literature are not feasible or, in any case, hardly effective when deployed in real devices. Therefore, further research is needed to better characterise the energy and hardware constraints of the devices and to develop more realistic models of the hardware architectures.

Existing Trends

The research community is well aware that solutions obtained on the basis of oversimplified mathematical models are often not feasible or inefficient in the real world. On the other hand, modeling is often the only way possible to test novel solutions without requiring the development of the entire system, which might be impossible because of cost and time constraints, or practical impairments (think, for instance, of emergency systems that are designed to operate in case of hearth-quakes or other catastrophes). Hence, there is a trade-off between accuracy and simplicity.

The research is, thus, following this trend. The aim is to develop more realistic models for those aspects that may greatly impact on system performance. For instance, a few studies have been recently focusing on the definition of models that describe the trajectory of vehicles in different areas (urban, rural, highway). Such models are tailored on data gathered from real-world measurements and, hence, simplified in order to provide a compromise between simplicity and accuracy. Similarly, studies on the depletion process of batteries have led to the definition of more-realistic energy-consumption models that can be exploited for the design of more effective energy-aware MAC and routing algorithms. The trend involves also other aspects, such as the asymmetry of coverage area, interaction among different aspects, calibration errors in device components and so on.

7.1.24 Experimentation

Description and Relevance

The analysis of the existing literature on wireless sensor networks and cooperating objects reveals a significant unbalance between the number of papers involving simulations and the articles describing significant experiments and applications with existing hardware. In fact, many existing methods have been validated only in simulation.

Moreover, many papers involving experimentation, only consider particular working conditions, which does not help in making their results statistically relevant.

The key issues are the efforts and cost required for the experimentation with wireless sensor networks and Cooperating Objects in the field, particularly when consider not only wireless sensor nodes but also the deployment of objects such as vehicles and machines.

The networks of Cooperating Objects interacting with the environment are very complex systems difficult to model mathematically due to the complexity of the interactions in real working conditions and the eventual dimensionality of the problem. Simplified models are usually adopted by neglecting effects that may have a significant influence in the performance and reliability.

Therefore, the real environmental working conditions may affect seriously the performance of wireless sensor nodes and their communication capabilities. The problem is more severe in mobile nodes in which the mobility creates additional interactions with the environment and poses communication issues.

The realisation of real-world experiments has to be supported by several tools that ease installation, testing, debugging and monitoring of the experiment. In deployments with many nodes, it is not feasible to connect each node to a computer to install the program. Especially for long running Cooperating Objects, the software on the deployed nodes has to be updated to correct errors or to add new features. A developer also likes to gather internal information of Cooperating Objects to assess the performance of algorithms. In case of errors, the developer needs a way to debug the application remotely if the cause cannot be determined only be looking at the monitored internal information.

Since available Cooperating Objects platforms differ in hardware such as processor, radio and sensors, an application can show very different runtime characteristics on another platform. Development tools should, therefore, provide development and prototyping support for different types of hardware in such a way that easy porting from one hardware platform to another is possible.

Existing Trends

The development of simulators based on relatively simple models has been the main approach to support the analysis and design. Statistical techniques are used to cope with the involved uncertainties and failures in the components of the networked systems. These simulation models are useful for a first validation of the theoretical work but in many cases are not enough to verify if the advances in research can be used in future applications.

In some cases, the information obtained in relatively simple hardware experiments is typically used to improve the models of the components. Thus, for example, experimental studies have been conducted to model the radiation pattern of the nodes antennas of wireless sensor networks. This information can be used in the simulations to analyse the network and also to develop new techniques such as for example localisation techniques in wireless sensor networks.

It should be noted that, the modeling of individual components is not enough. In fact the analysis of the distributed system requires the study of the cross effects and the propagation of these effects through the networked system. This analysis is more difficult when considering sensor-actuator networks because the actuators can modify the environment generating effects that could be difficult to predict. Moreover, an important problem is that complex models typically involve many parameters that are difficult to obtain. Then, experiments using real hardware are required for validation.

The next step is the design of laboratory experiments useful to validate some subsystems using real hardware. A well known design methodology to analyse the performance before the real implementation in the field is the "hardware in the loop" approach. This approach is useful to address first implementation issues of the involved hardware, but it is not enough because the limitations of the software models connected to the hardware. In some cases it is also possible to use laboratory pilot plants, small equipment, and hardware scale models, such as small vehicles. Then, it is possible to implement distributed sensing and control approaches by using, for example, nodes of wireless sensor networks and/or a team of these small vehicles in the lab involving suitable communications and wireless networking. These experiments provide significant validations for many applications but are not enough in others due to the impossibility to have realistic laboratory scale environments.

The only solution to validate networks of cooperating objects is then to perform full experiments in the field. In some applications, such as home applications, the required scenarios are relatively easy to be conditioned. However, in others, such as disaster management or civil security, field experimentation in realistic conditions is extremely difficult. Then, important efforts have to be devoted to design realistic scenarios capturing the main characteristics of the real applications. The information obtained in these experiments is of paramount interest to improve the design, following a bottom-up approach that could complement the top-down design from the application requirements. Furthermore, the organisation of experiments in these scenarios is very costly and involves important logistic requirements. Then, significant human resources and funding are necessary.

Developer support for hardware experiments is still limited concerning monitoring and debugging. Monitoring solutions are usually designed only for a specific application, but general and integrated solutions are missing. Wireless sensor nodes are usually equipped with less powerful processors that do not have debugging capabilities without support from external hardware. In testbeds, it is possible to connect each node to a wired or wireless infrastructure over which the debugging of the node is conducted. Such an infrastructure also allows for the remote installation of new code. Several software solutions exist as well that work on small Cooperating Objects without the need for additional hardware.

7.1.25 Social Issues

Description and Relevance

Small cooperating objects like sensor networks are being developed to monitor the environment unobtrusively. In some scenarios, e.g. monitoring of small animals, this is essential. The same technique can be used to spy on people without their knowledge. One of the key gaps is how to prevent this illegitimate use.

Huge databases can be built to create detailed profiles of people. Nowadays, everybody leaves already many traces, e.g. when paying with credit card, but these traces are fragmentary. If things of daily use are part of Cooperating Objects or if Cooperating Objects are able to monitor just the behaviour of people, the information on an individual becomes much more continuous and comprehensive.

Such information is very valuable to companies since it helps to categorise people. Advertising can be tailored – which might also be beneficial for the people –, but they can suffer detriment, e.g. when a salesperson knows instantly that the person entering the shop does not have much money.

People show an unspecific fear of loosing control over personal property if the technology is invisible or imperceptible. They are also afraid of being tracked or patronised by this technology. Another fear is a “big brother” knowing everything about the people whereas they do not know anything about who knows what. If these social issues are not solved it is very questionable whether or not Cooperating Objects will find broad acceptance.

The security features described in section 7.1.16 only help to prevent the misuse of a Cooperating Object without the knowledge of the owner. But they do not solve the privacy problem if a Cooperating Object is deliberately installed to spy on people.

Existing Trends

Since RFID tags are a technology ready for and used in the market most solutions exist for them. RFID tags are developed whose information can be deleted electronically by a kill command. Also, hardware solutions have been proposed, e.g. antennas that can be rubbed or broken off, thus making it impossible to read out the RFID tags. But this is only possible for tags people can control. It does not prevent the use of unknown tags or spying sensor networks.

7.2 Timeline

The gaps described in the last sections only identify the type of work that needs to be done in certain areas, but do not give any indication as to when they are expected to be solved. Therefore, we have arranged the previous gaps along a diagram that shows the approximate time when each gap starts, ends and is expected to have a major breakthrough. This timeline reflects the estimation of the `Embedded WiSeNts` consortium based on our in-depth research in the studies, visions, gaps and trends, also taking into account the dependencies between the gaps as shown in the last sections. It is important to mention that our estimation assumes that research is driven by the current needs of the users and that no research direction is promoted due to, for example, this roadmap. Thus, the timeline does not show the ideal state. But this process made it possible to extract the predominant research areas afterwards (see section 8).

The timeline contained in this section is divided into three zones that correspond to the short-term, medium-term and long-term solutions covered previously in other aspects of this document. As already mentioned there, we define short-term starting today and ending in 5 years, medium-term between 5 and 10 years and long-term between 10 and 15 years.

Additionally, we show each gap using the notion of a deltoid, which is an easy-to-draw geometric figure that represents Rogers' well-known curve of technology adoption [54]. In this bell-curve, shown in Figure 7.1, there are a number of innovators and early adopters that drive the technology and are not afraid to use it. Then, the early and late majority are convinced of the new technology and finally, the laggards start using it. Somewhere between the early and late majority there is usually a major breakthrough that drives the late adopters into the use of this new technology.

In our particular case, the deltoids in this section show the adoption of certain technology and the breakthroughs that might happen regarding research. Moreover, we do not believe

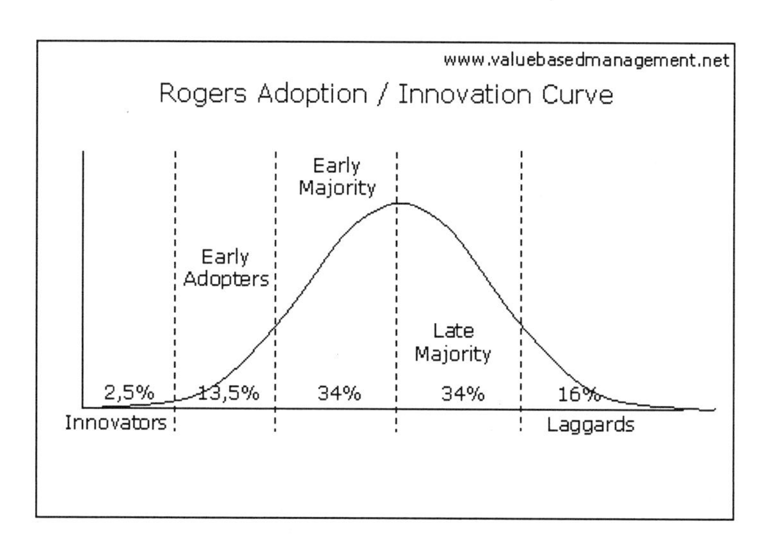

Figure 7.1: Roger's Innovation Curve

that all deltoids (bell-curves) look exactly the same and have their point of inflexion at exactly the same position. Therefore, the deltoid representation is to be interpreted as follows:

Starting Point: This indicates the approximate point in time where research will start looking in to the gap in more depth. It is obvious that for each gap, there could be research groups or institutions that are looking at these issues before the starting point we have in the timeline, but we have neglected them in the figure.

Maximum Height: This is the point in time where we expect a major breakthrough in research to happen. There are two possible (and complementary) consequences:

- Research in this area starts to decay, since the main problem has been solved. There is still some work to do, but the research community agrees that they should start looking into other problems.
- The industrialisation and commercialisation of technology is boosted and, if the appropriate research problems are solved by then, commercialisation of Cooperating Objects in general will dramatically increase.

End Point: This indicates the approximate point in time where the research community will stop looking at this particular issue. As for the case of the starting point, there might be certain isolated groups that still do research on it, but the bulk of the research community is expected to move onto other topics.

Figure 7.2 shows the gaps explained in the previous section using this notion of a deltoid. Looking at the general picture, we can see the largest number of breakthroughs to happen in the middle-term, that is, between 5 and 10 years, which agrees with the estimation obtained from our survey regarding the point in time where Cooperating Objects will start to be used widely in the industry (see section 6.2).

As we have already mentioned, most of the gaps depicted in figure 7.2 are already being worked on in some form or another. Only two of the gaps mentioned in the previous section are expected to start 5 years from now: *Real-time* and *Social Issues*. The first one because of the nature of the problem and the second one because social issues will only arise as soon as the early adopters (especially from industry) start introducing Cooperating Objects more aggressively in our everyday lives. In general, gaps that are not being investigated yet or that need investigation throughout the predicted period seem to be the most promising lines of research.

Let us now look at each one of the groups of gaps mentioned in previous sections in more detail.

7.2.1 Hardware

We expect *Sensor Calibration* to be solved relatively soon in comparison to other gaps because unless this issue is solved in a satisfactory way, it is hard that sensors can be used in environments where costs play a major role, such as in the *Home and Office* domain. Other more industrial domains are willing to pay higher prices and, therefore, more sophisticated methods for sensor calibration can be used.

A similar argumentation can be used with the *Power Efficiency* gap. We expect a major breakthrough in a short to medium term, because of the importance of this issue for the adoption of technology. It is clear that this will remain an issue that needs to be investigated further in the future, but unless we are able to provide good solutions in a relatively short period of time, there might be a decrease in the interest of Cooperating Object technology.

Energy Harvesting, on the other hand, is a very hard problem that will require more time to find solutions that could be used on a more widely basis. And it looks like the need for more efficient materials that transform other types of energy into electrical impulses will remain an issue in the long term.

The same happens with the *New Sensor and Low-Cost Devices* gap. There are quite a few research groups, research institutions and companies trying to produce cheaper and cheaper devices, and this will continue throughout the evaluated period since the need for new and improved devices will increase with the adoption of technology. The addition of new application domains implies not only the creation of new devices, but also changes in the requirements of already existing ones that will have to be improved accordingly.

Finally, *Miniaturisation* goes hand-in-hand with cost and power-efficiency and we believe that the need for smaller and smaller devices will continue throughout the studied period.

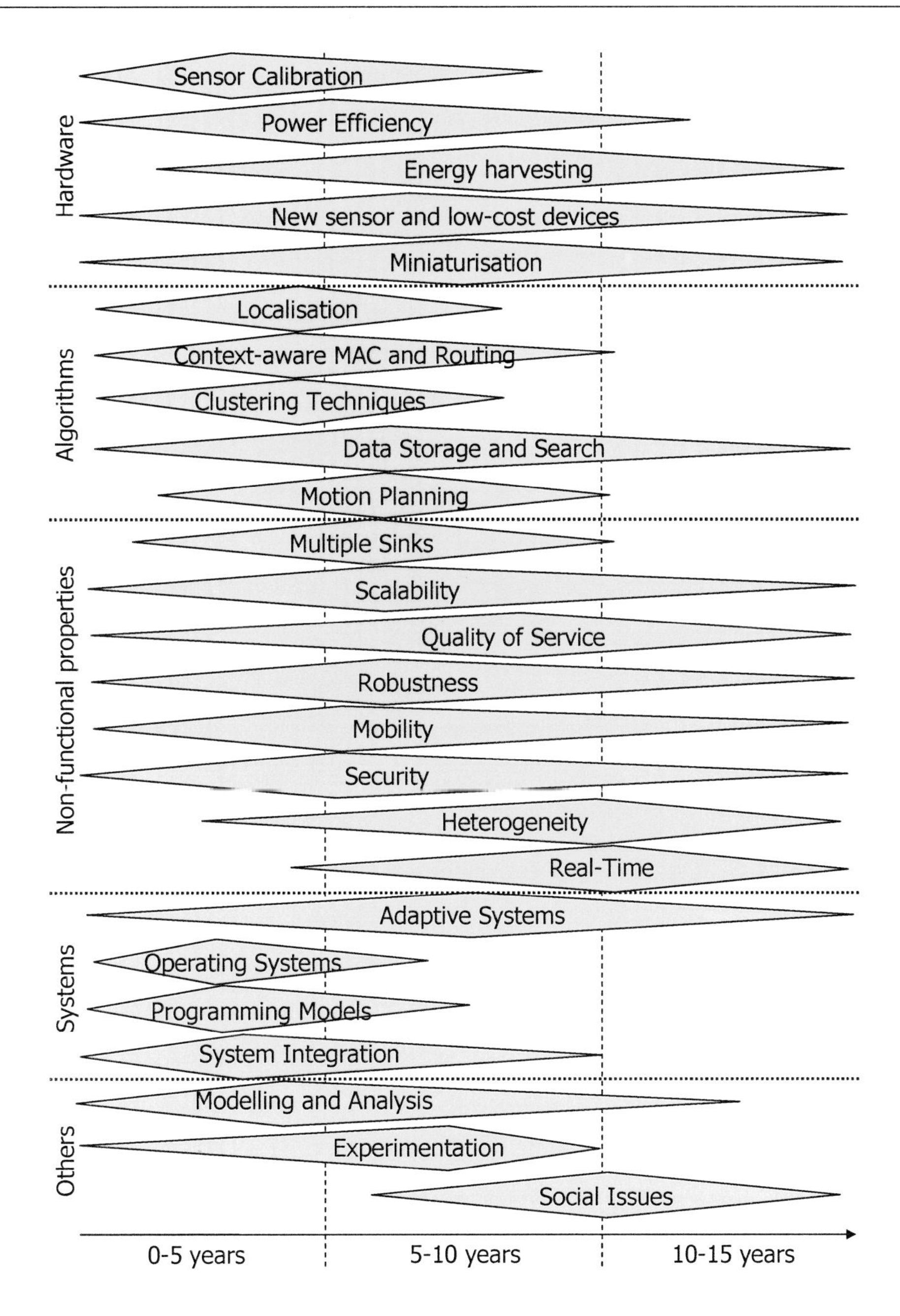

Figure 7.2: Timelines for Gaps

In the end, there might be even a fusion of different research areas, such as nanotechnology, that attempt to create even smaller devices than the ones envisioned for the field of Cooperating Objects.

7.2.2 Algorithms

With respect to algorithms, *Localisation*, *Context-aware MAC and Routing* and *Clustering Techniques* are expected to be well understood in the short to medium term, so that major breakthroughs in all areas can happen at that time. These are research topics that have received quite some attention in the past years and that will benefit from several more. At the same time, there algorithms are so fundamental for the adoption of Cooperating Object technology that a slower development would definitely affect their acceptance.

We also expect other algorithms such as *Data Storage and Search* to require longer and to continue to be relevant in the long term. The main reason for this is the fact that requirements of applications will involve the design of new types of solutions that, due to the data-centric nature of Cooperating Objects, will mostly affect the design and implementation of data storage and search algorithms.

Finally, *Motion Planning* is expected to lie somewhere in between, since it requires good localisation techniques that work in heterogeneous environments. However, in the medium-term, we expect to have good planning algorithms that combine the benefits of current robotics research and the use of Wireless Sensor Network technology.

7.2.3 Non-functional Properties

Non-functional properties, in general, will remain a hot research topic throughout the studied period and will continue to drive the development of new algorithms and system characteristics for applications based on Cooperating Objects.

More specifically, we expect the issue of *Multiple Sinks* to be solved first, since it is a natural extension of the kinds of algorithms and optimisation procedures available today, but other non-function properties such as *Scalability*, *Robustness*, *Mobility*, *Security* and, in general, *Quality of Service* will receive some attention well after then issue of multiple sinks is solved. Especially with the increase in application-driven research and algorithm implementation, non-functional properties play a crucial role that distinguishes an application designed for a specific environment (or application domain) from another one.

The issue of *Heterogeneity* will probably be worked on mostly in the long-term, since this issue requires efficient and fault-tolerant algorithms for each of the parts they contain. The fact that there might be different parts of the network cannot work efficiently together unless the individual and homogeneous parts do internally.

Finally, the issue of *Real-time* is a very hard one since there are a series of assumptions in Cooperating Object technologies that make real-time a very challenging problem. The use of wireless technology, for example, will probably require a redefinition of real-time in such a way that certain guarantees can be achieved, but this seems today like a daunting task that will require years of further research.

7.2.4 Systems

We expect the issues related to *Operating Systems*, *Programming Models* and *System Integration* to be solved relatively soon, since they are the basis that other algorithms, etc. use for their work. Without a clear view on these issues, the adoption of Cooperating Object technology might suffer.

On the other hand, *Adaptive Systems*, that is, middleware solutions that use the underlying operating system and programming abstractions to provide additional functionality that deals with adaptation, are expected to remain a hot topic of research throughout the studied period. The reason for this is the need for optimisation of applications independently of the environment they are immersed in. This can be achieved by the use of adaptive systems that need to be worked on in order to be able to work on newer application domains that will arise as a result of a wider adoption of Cooperating Object technology.

7.2.5 Others

The issue of *Modeling and Analysis* has started to receive a lot of attention and we expect a major breakthrough in the short to medium term. It is clear that if the application domains change and their requirements need to be mapped to mathematical models, this will require some work, which is why this issue extends well into the long term area. However, good modeling and analysis tools are crucial for the adoption of Cooperating Object technology.

Experimentation, on the other hand, needs good modeling tools in order to create the appropriate test cases and, therefore, we expect a breakthrough for this particular gap after the modeling and analysis issues have been well understood.

Finally, *Social Issues* cannot be worked on until most of the other areas and gaps have matured to a level that makes Cooperating Object technology interesting for a wider audience. In order to solve these issues, it will be necessary to introduce new regulations and laws that control the use of Cooperating Object technology. We expect this to happen in the long term.

7.3 Market Prediction

Based on the timeline and the description of the gaps of previous sections, we provide an analysis of the expected evolution and market performance of the application areas presented in section 4.1. The prediction is based on the information obtained through the

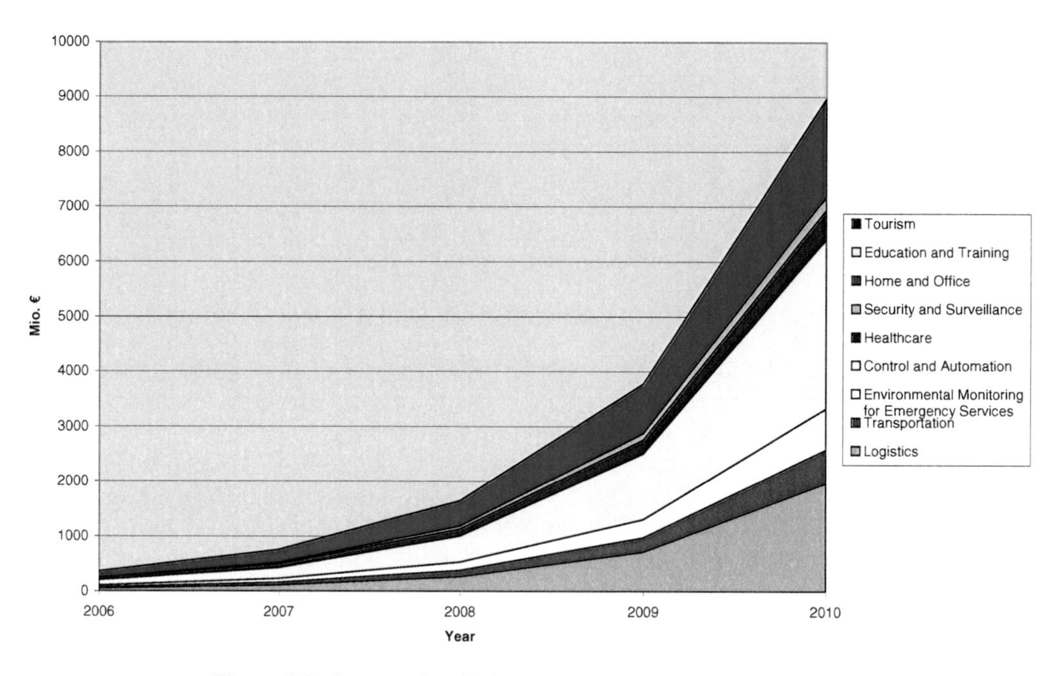

Figure 7.3: Cooperating Objects Market Prediction 2006-2010

ON World Inc. studies [50] using the aforementioned mapping of application domains, and data from several reports from Frost & Sullivan [25, 23, 24].

The growth rates used in this prediction uses the rating of application relevance we obtained from our own survey (explained in section 6.2), the gaps for the application areas and their estimated time for a major breakthrough. For the ease of exposition, we will consider the two periods 2006-2010 and 2011-2015, which correspond to the short and medium-term periods mentioned in previous sections.

Both graphs show the revenue in thousands of dollars we expect these markets to generate. These numbers have been obtained by taking the estimates from the ON World Inc. studies on the revenue values for each application domain and estimating the growth rate based on our timeline. The computed prices per deployed node of these studies match the expected amount customers in these sectors are probably willing to pay for one node.

Looking at figure 7.3, the areas *Logistics* and *Control and Automation* are expected to grow fastest. Their peak is expected to be in 3–4 years since the most important gaps required for their development (localisation and security) will be solved in the short term. On the other hand, *Environmental Monitoring* grows slightly more slowly, but also with a peak after 4–5 years following the aforementioned timeline. All other areas are expected to

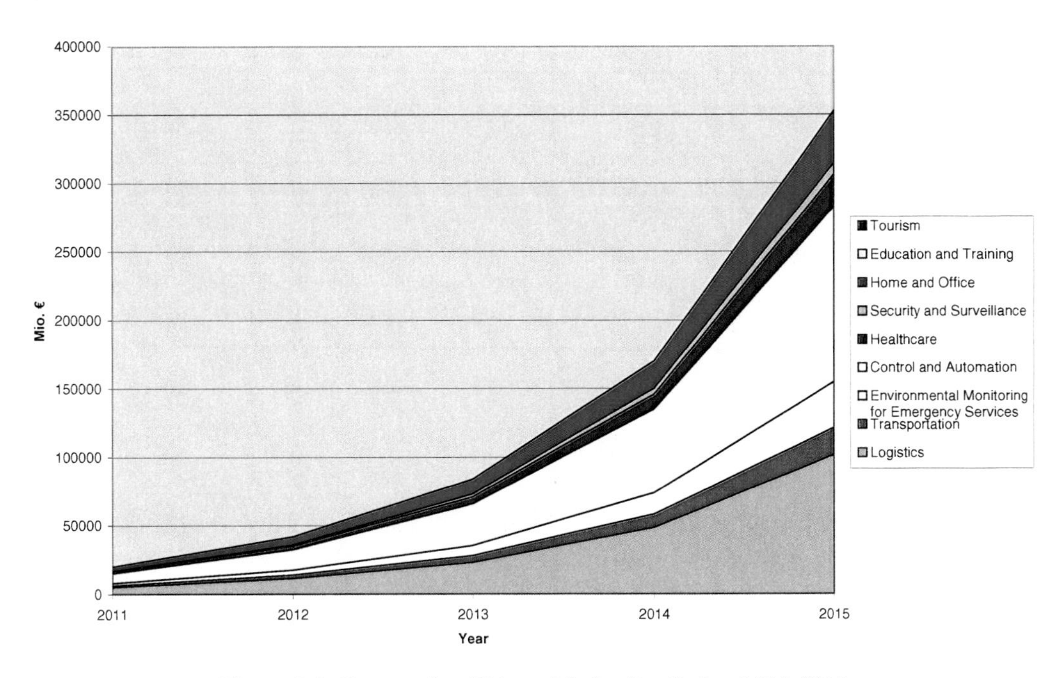

Figure 7.4: Cooperating Objects Market Prediction 2011-2015

grow more slowly since the solution for their requirements will probably take longer, especially application domains that require real-time characteristics, complex data-management solutions, energy harvesting and quality of service.

As opposed to the ON World Inc. study, we do not expect the *Home and Office* application domain to grow as fast since, in our opinion, there are quite a few social issues that need to be addressed before a considerable adoption in this application domain happens.

Our prediction for the years 2010-2015 is depicted in figure 7.4. This graph shows several differences when compared to the graph of years 2006-2010 that can be explained as follows:

We expect the *Logistics* and *Control and Automation* domains to lose some momentum and slow down with respect to the growth of the first 5 years. The reason for this is that we expect a certain market saturation that will imply the growth of other areas. Additionally, at the end of this period we expect major breakthroughs in the following areas: mobility, scalability, real-time, etc. which will allow for the steadier growth of other areas such as *Transportation* and *Healthcare*.

7.4 Technology Inhibitors

The market predictions presented in the previous section assume that there are no real technological showstoppers or inhibitors that drastically change our views regarding the major breakthroughs and adoption of Cooperating Object technology. However, this is a very optimistic view, and the identification of potential inhibitors should not be neglected.

For this purpose, we rely on two sources of information that deal with this issue. The first one is the ON World Inc. studies that indicates clearly what the biggest technological inhibitors might be, and the second source of information is our own survey, conducted as part of the "From RFID to the Internet of Things" workshop in Brussels.

The following two sections give details about the results.

7.4.1 ON World Inc. Studies

ON World Inc. carried out a survey with 58 OEMs and platform providers in residential, commercial buildings and industrial markets and published it in the report "Wireless Sensor Networks – Growing Markets, Accelerating Demands" from July 2005.

Among the results provided, one of the most interesting ones for the purposes of this roadmap are the reasons for late adoption of wireless sensor technology, depicted in figure 7.5.

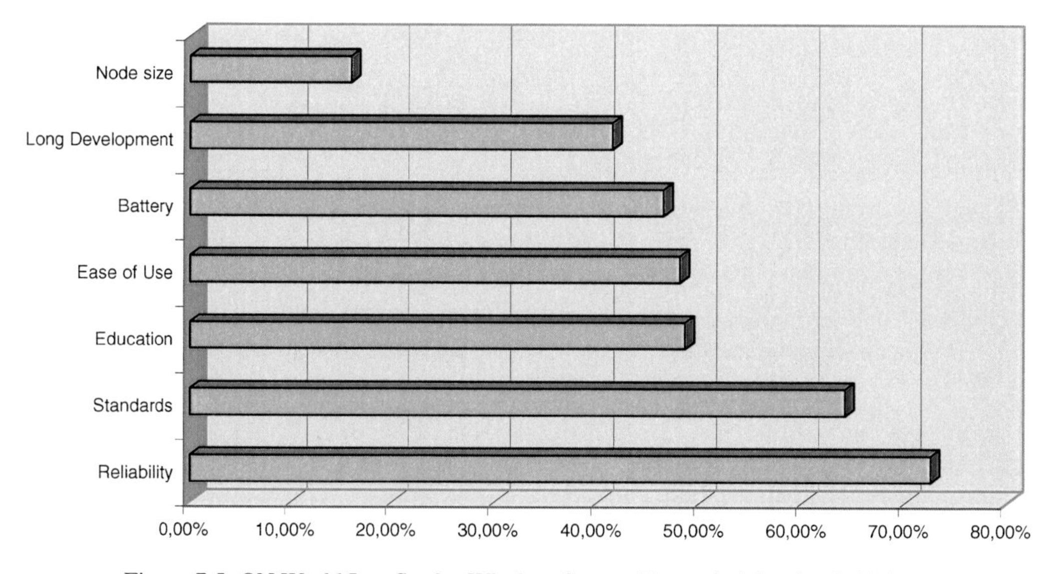

Figure 7.5: ON World Inc. Study: Wireless Sensor Network Adoption Inhibitors

It is interesting to notice that reliability scored highest for all market segments. For industrial monitoring applications, reliability is definitely one of the most wanted properties and lack of it is considered a deterrent, since the value of the monitored machines and goods can be very expensive. For other market segments, the high rating of *reliability* can result of the fact that the participants of the survey have just started to deal with sensor networks and are, therefore, a bit concerned when comparing wireless with wired communication. ON World Inc. also predicts that mesh networks, which provide the best solution for industrial-strength reliability, will make up 90 percent of all Wireless Sensor Networks by 2010.

Standardisation is the second most important barrier since many companies insist on interoperability. It is predicted by ON World Inc. that in 2010 ZigBee or ZigBee variations will make up 64 percent of all deployed nodes compared to 10 percent in 2006. Thus, the increasing use of ZigBee for all sorts of Wireless Sensor Networks can mitigate this potential inhibitor.

Although *Education* of the users about the benefits of Wireless Sensor Networks is place 3 in the overall result, it is the major concern for the residential control and automation sector. The difference to the other market sectors is that many users have to be convinced instead of only a few to generate the same sales volume.

The ON World Inc. study "Wireless Sensor Networks Technology Dynamics" from July 2005 lists "node costs" as another major roadblock. Especially for residential control and advanced metering, the monitored components are of low value but occur in large numbers. Therefore, nodes should have prices that make them affordable for the average type of customers of each sector.

7.4.2 WiSeNts Survey

Besides the application rating presented in section 6.2, the participants of the survey carried out by the `Embedded WiSeNts` consortium were asked to rate how the immaturity of the vertical functions identified in section 4.3 slow down the industrial deployment of the application area they work in. The results are shown in figure 7.6.

It is interesting that, according to this survey, hardware and lower layer technology like operating systems, communication or time synchronisation seems to be a minor problem. Among people dealing with hardware development, hardware platforms received a higher rank than among all people. In contrast, all functionality related to data like storage, search, aggregation or consistency are rated as bigger roadblocks. The biggest problem, however, is still security, privacy and trust.

This trend was also confirmed by a free text question in the same questionnaire, where we asked the participants about the main market entry barriers for Cooperating Objects. The most frequent answer had to do with security and privacy issues, followed by pub-

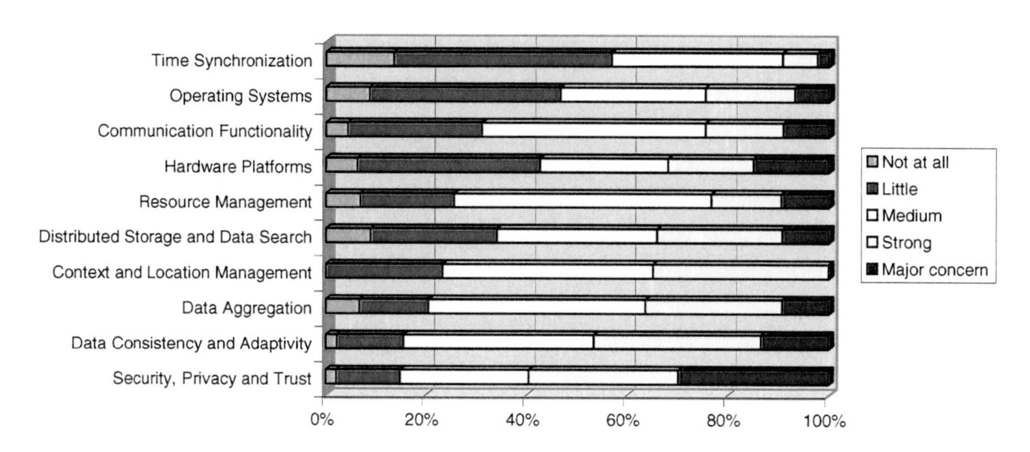

Figure 7.6: Technological Roadblocks of Industrial Development

lic acceptance or, in the words used throughout this document, social acceptance of this technology.

Another problem was the lack of convincing applications for Cooperating Objects and, therefore, still no convincing business case that might justify bigger investments from the industry. Deficiencies in technology and the need for research follows afterwards. Finally, the issue of standardisation, or lack thereof, was considered one of the major inhibitors and, interestingly enough, only 10% of the participants in our survey indicated costs as a major barrier for the acceptance of Cooperating Objects in industry.

8 PREDOMINANT WORK AREAS

After discussing the research gaps, timeline for their development, market predictions and potential inhibitors, it is necessary to determine the predominant work areas and how these areas could be addressed by the major players, let them be the research community, industry, the end-user, etc.

The gaps identified by the `Embedded WiSeNts` consortium show the view of developers working directly on software for Cooperating Objects while the studies of ON World Inc. and the survey carried out by `Embedded WiSeNts` represent the view of manufactures and products developers which do not deal on a daily basis with Cooperating Object technology. Moreover, the studies from ON World Inc. refer mostly to application scenarios where Cooperating Objects are already in use or are about to be used, while our own analysis also shows further probable application areas.

In some respects, the view of developers and manufacturers differs slightly, e.g. concerning the actual algorithms that make up an application and, therefore, future research projects will have to address application-driven issues in order to avoid diverging too much from the expectations of industry.

In this chapter, we propose a series of work areas and items (section 8.1) and, in section 8.2, a possible way to organise these activities in such a way that all major players are sufficiently involved. We also identify some potential roadblocks in section 8.3 and conclude this chapter with some final recommendations for future research programs within the EU in section 8.4.

8.1 Work Areas

It is interesting to notice that several issues have been identified by the research community, the industry and end-users as important work items. This section presents the most important issues that should be considered in the future. These were determined by selecting gaps that either require long-term research or are expected to start late but should start earlier for the vision of Cooperating Objects to become a reality. The selection of the areas also took into account the market potential of different application areas and the market entry inhibitors identified by the ON World Inc. study and our own survey. In order to push Cooperating Objects on the whole, the work areas are selected to cover a large amount of Cooperating Object applications. There might be research areas that are essential for some applications but are of lesser interest for the whole Cooperating Object community. However, this should not result in general purpose solutions only since – as said before – we encourage application-driven research. It is also important to mention that the gaps identified in section 7.1 are still important even if they have not made it to a predominant work area.

For the ease of exposition, we have identified three different types of actions: actions related to research, actions related to educational activities and other actions.

8.1.1 Research Actions

Regarding research, there are five major topics that should be addressed and that summarise some of the major gaps described in the previous chapter.

Data Management Algorithms

The handling of large amounts of collected data is a promising research area, including distributed multi-sensor environment perception. This includes a variety of single techniques that should be combined in the appropriate way to create optimised solutions. For example, data has to be aggregated in order to reduce the amount of data to be transmitted. At the same time, its storage within the network is not an easy task that should be addressed, as well as the methods to search and query it. To improve on the availability of data, it is possible to use replication techniques, but this might lead to consistency problems that should be solved in an energy-efficient way.

In general, the goal of the data management techniques is to simplify the usage of Cooperating Objects. Especially manufacturers are concerned about the integration of Cooperating Objects into existing software. Therefore, they need an easy and intuitive interface to get the interesting data without knowing the internals of a Cooperating Object.

Hardware

Hardware which is small in size, consumes little power and does not cost too much is not an issue for all application areas, but viable for others to become relevant. For example, if everyday devices should become Cooperating Objects to be included in a Home Automation and Control application, hardware must be cheap, since it will be applied in large numbers, and small to be included and integrated into small devices as well. Moreover, it is not feasible to change batteries in hundreds of objects every few months. Therefore, the issue of power consumption at the hardware level plays a very important role.

Miniaturisation of hardware and its standardisation will lead to the broad adoption of Cooperating Objects, and consequently the price per node should drop significantly. Additionally, the search for alternative energy sources for Cooperating Objects is a necessary research area since the requirements of applications in terms of the frequency of data collection, the amount of data collected, etc. will increase over time.

Non-functional Properties

Non-functional properties are not directly implemented as algorithms but influence their design. Many algorithms have been developed that are said to be scalable and to be able to work well in mobile environments. However, there is no real classification of scalability and mobility. Thus, one scalable algorithm supports only a few hundred nodes while the other works with tens of thousands. Moreover, a single application area might differ significantly from another one and its requirements regarding non-functional properties might not hold in the new environment. Therefore, algorithms should concentrate on a few properties that are motivated well by real application scenarios.

The same also applies to Quality of Service concepts. The difference to the aforementioned properties is that they are given by the environment while the required quality of service is given by the user and can change over time. A detailed study of the application areas about these requirements is missing. Most of the time, developers tend to design an algorithm which gives "best" quality of service while neglecting that it can be useful to reduce the quality of service requirements to improve other properties, e.g. power consumption.

"Security" as a property can be considered a type of quality of service, but should be considered explicitly. All application scenarios will require a certain level of security if they evolve from the prototype state. For control applications, it is obvious that an intruded can cause damage to the controlled objects. But also for simple monitoring applications, a secure network will increase the trust of people in technology. Since an existing system cannot be secured afterwards but has to be designed with security in mind – which is usually not the case –, new research projects have to consider security aspects from the very beginning as a cross-layer task.

Adaptive Systems

The vision of Cooperating Objects that are easy to use, easy to combine with existing software, fast to develop and working on standard platforms can become reality if there is an construction kit with simple building blocks that can be just plugged together. These blocks will be targeted to a small set of classes of system and non-functional properties. To be efficient at the same time, a Cooperating Object has to adapt with respect to these properties.

Adaptation is normally supported by a middleware or some adaptive system software. In the best cases, the end user is not even aware that adaptation is happening and it is not necessary to specify in what cases which algorithm should be used. Rather, the end user can describe in a high-level manner how the properties are linked and the underlying system takes care of the rest.

Additionally, there is a need for the development of techniques and methods to enable the autonomous cooperation of many (possibly moving) objects in order to fulfill a given task. These tasks might include the allocation and distribution of work as well as distributed execution of control tasks.

System Integration

One of the biggest problems when dealing with real-world scenarios in distributed systems is the task of system integration. Even when most individual pieces work properly, their combination into a single system might trigger the detection of problems that have not been discovered before.

There are two types of system integration that need to be performed: on the one hand, the integration of a testbed in the lab and, on the other hand, the integration of a complete system into a (possibly hostile) environment.

For testbeds, the development of new applications using a full-featured IDE, the easy installation of the application on the nodes of the testbed, systematic testing, run-time monitoring and debugging – also for big testbeds – are most important.

Since the final deployment of Cooperating Object technology does not usually happen in a controlled environment like in testbeds, the original developers of the product cannot foresee its exact characteristics. Cooperating Objects have to be connected via transit networks – wired and wireless, from a defined topology to Mesh networks – to end systems. Since the types and number of Cooperating Objects expected in the future is so high, the interoperability of independently developed systems plays a crucial role for the acceptance of Cooperating Object technology.

These issues have to do not only with the physical integration of components into a single system, but also with the software integration and the standardisation of APIs, hardware interfaces, etc. so that the end-user can really benefit from a variety of software and hardware manufacturers.

8.1.2 Educational Actions

In terms of education, it is expected that Cooperating Object technology will be a part of our daily lives. Therefore, there is a need for education of the industry and of end-users in order to promote the new technology and to allow for potential users and customers to understand the benefits and risks associated with the use of Cooperating Objects.

At the same time, companies that would like to enter this market need to be learn about the potential and the power of this new technology so that they can properly address the needs of their customers. This is especially true in application domains such as *Tourism*, *Education and Training* or *Home and Office*, where the end-user depends on the knowledge and support from the companies they buy their products from.

Additionally, end-users have to be taught about Cooperating Objects to increase public acceptance. For this purpose, legislative regulations might also help, especially if they deal with some of the aspects that most interest the public opinion. issues such as privacy for the use of Cooperating Objects have to be solved before a wider adoption of Cooperating Objects is to be expected. The lesson learned from RFID is that technology introduced silently results in bad reputation of both, the technology itself and the companies that promote it. Therefore, research projects should cover security and privacy issues theoretically, in the form of algorithms, and by including end-users in a public test of their prototypes.

8.1.3 Other Actions

Under the "Other Actions" heading, we have included other types of activities that do not directly affect research or the education of individuals or companies. In this respect, the most prominent action that needs support from all major players is the process of standardising the hardware and software available for Cooperating Objects.

Standardisation

After a phase of research where different hardware platforms, radio stacks, and protocols have been proposed, there is a need for a consolidation and standardisation phase that drives the adoption of this new technology.

When discussing this issue with the industry, many manufacturers are still waiting for standards and standardisation committees to indicate which technologies should be considered in more detail and, in a sense, show them the way that will be favoured by most competitors and potential suppliers. This also affects their prices and will definitely benefit from hardware and software standards. On the other hand, standardisation committees composed by the major industry players should be formed in order to promote this technology. One of the first attempts is the standardisation of ZigBee, but other issues, not just communication should also be addressed. For example, IEEE 1451 should be extended and adapted to Cooperating Objects and Wireless Sensor Networks.

Developers and researchers can also benefit from the use of standards, since they can concentrate on only a few platforms, communication technologies, operating systems and algorithms. Therefore, the development of standard platforms, software, interfaces and tools for debugging and development should be favoured over the development of more proprietary forms of hardware and software.

8.2 Organisation of Activities

In order to coordinate the activities described in the previous section and to make clear the connections among them, it is important to clearly state the key players that should be involved in this process. This will help us organise future activities and make sure the right entities are involved.

Future research on Cooperating Objects is determined and influenced mainly by three groups: the research community, the industry and the end-users. The research community consists of all academic and industrial researchers actively doing research in areas related to Cooperating Objects, which might be enabling technologies, the gaps identified previously or other technologies that contribute to the success of Cooperating Object technology. For the most part, they do not manufacture Cooperating Object products for large numbers or for end-users and are mostly concerned with implementing prototypes that show the capabilities of new technologies.

The industry group consists both of OEMs and suppliers of Cooperating Object technology that either manufacture it for the end-users or for other companies that might incorporate them in products for the end-user. Both types of companies require certain knowledge about Cooperating Object technology in order to use them in their daily activities.

Finally, the end-user group consists of all private users who buy Cooperating Object-based products and support from industrial manufacturers. The main difference between industrial and private end-users lies in the fact that the industry has a strong financial interest in the use of Cooperating Object products.

The interactions between these three groups is depicted in figure 8.1 and is described in more detail below. In the figure, the circles indicate the major players we have just discussed and the arrow represent interactions between them in the form of the work areas explicitly stated in the previous section.

Regarding the activities of the three groups, the first type is the ones performed within a certain group. Standardisation activities and the development of algorithms, hardware, energy harvesting methods and system integration activities belong to this group. The standardisation part involves (and has to be driven by) the industry, whereas the rest are mostly part of the research activities performed in universities and research institutions.

Although not explicitly depicted, we assume that interactions between different members of the research community or the industry are necessary. For example, cooperations between hardware and software specialists is definitely needed for the development of efficient energy-harvesting solutions. The same applies to the industry and to the standardisation procedure. This activity can only work if there are key players from the right industries.

Regarding interactions among the groups, the end-users need to provide the industry and the research community with enough information as to know what kind of application requirements they have. Usually, the end-users give feedback to the industry by buying a product or not. Market research, surveys or involvement of end users in prototype test

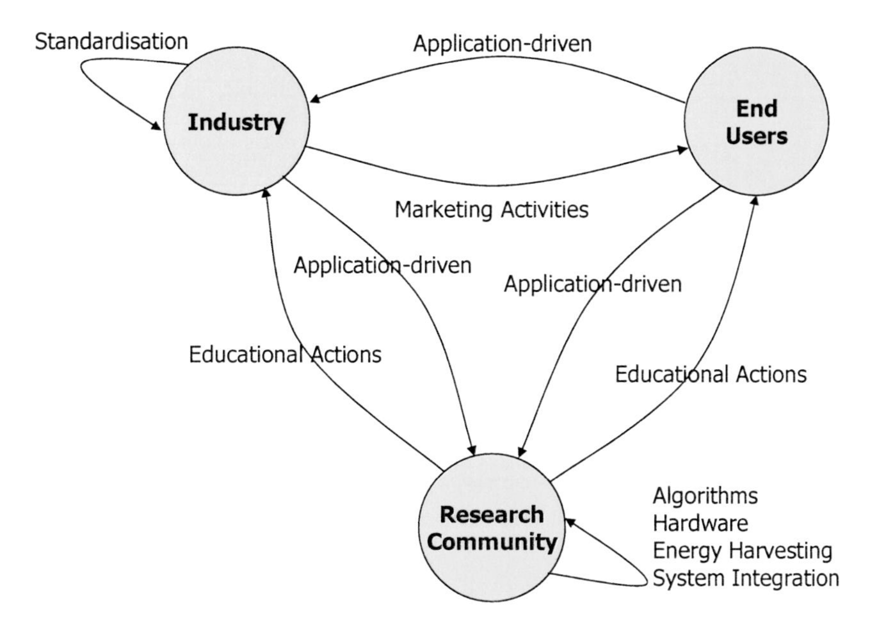

Figure 8.1: Main Players of Research and their Interactions

scenarios other possibilities to consider user interests. This way, application-driven research and commercialisation can be performed in a way that fits the real needs of end-users. This is especially important for the research community that more often than not has little connection to the needs of end-users or the industry.

Cooperation between the research community and the other groups can also be performed in the other direction. There is definitely a need for education of end-users and industrial partners in the following two ways: first, the awareness of Cooperating Object technology and, more specifically, its power to be increased in the industry. Having this knowledge, the industry is able to develop Cooperating Object visions or products and give feedback to the research community. In fact, it is only with a bidirectional interaction that the industry can cooperate and benefit from research properly. The input from the research community will also help the industry to define robust and more future-proof standards. Secondly, end-users need to learn the benefits of using the new technology and the pros and cons of Cooperating Objects should be openly discussed in order to obtain a level of awareness at the end-user level that drives the industry and pure research. As an example, one of the activities that could be performed in conjunction with users are legislative and social awareness actions.

Finally, marketing activities between the industry and the end-users are different from the education interaction between the research community and the end-users. Concrete products have to be shown to the end-users for them to understand the benefits of using Cooperating Object technology. Users also have to be taught how to use it and how to protect themselves from potential misuses, not in an abstract way, as the way the research community would teach those issues, but in very practical ways with concrete products. In return, the industry obtains from end-users related groups like consumer advocates information as to how to improve their products.

As can be seen from the described interactions, the main part of research should be driven by applications. It is, therefore, necessary to build real systems that the users can look at, interact with and improve. Research projects should also concentrate on rather small, well defined applications in order to generate prototypes that can be tested in real-world scenarios. Starting from these specific cases, general solutions and answers should be provided.

8.3 Potential Roadblocks

Working with members of the industry and with the participants of the `Embedded WiSeNts` survey we have described in previous sections, it seems clear that, although the opinion of all experts indicates that Cooperating Object technology has clear chances of success, there is always the possibility of failure if certain issues are not solved properly or in a timely manner.

In this section, we describe the most important potential roadblocks that have appeared during these conversations and interactions, although it is very hard to really pinpoint their severity in such a way as to quantify their effect in the gap timeline or the market predictions.

8.3.1 No Clear Business Models

One of the main potential roadblocks for the adoption of any kind of technology is the lack of a business model that supports it.

Some of the industrial partners we have talked to during the writing of this roadmap have expressed their concerns regarding the wide adoption of this kind of technology and, although there might be niche markets that use Cooperating Objects, they are still waiting for the "killer-application" that justifies bigger investments, especially when research is pushing for systems that should be scalable, provide a certain level of quality of service, etc. But to create a business model, a trade-off needs to be done between usage, complexity, cost and revenues.

On the other hand, there are more and more business models that use Wireless Sensor Network technology (one of the canonical examples of Cooperating Objects we have discussed throughout this document) in such a way that it would indicate the need for such systems. Nevertheless, they often cite the case of Mobile Ad-hoc Networks (MANETs) as a very interesting technology in search of a market.

Cooperating Object applications could be "support" tools for daily activities, e.g. healthcare or logistics, but they are really useful only if applied on already adopted or existing procedures. However, in this case, Cooperating Object technology should provide additional advantages and not just be a replacement for legacy (usually wired) systems.

For the case of Cooperating Objects, it is probably too early to determine whether or not strong enough business models will appear and, as far as early adopters are concerned, there are enough examples of companies that make their living nowadays selling technology that can and will be used in this field. However, it might be necessary to work tightly with end-users to identify the real needs and, therefore, business models with high potential.

8.3.2 Lack of Standards

Independently of whether or not there are business models for Cooperating Object technology, there is a clear need for standardisation in the field. With the creation of a new field, it is obvious that early adopters need to provide a pragmatic solution in order to "show something that works", but after a certain time, the industry needs to come together and agree on a common ground for future developments.

The potential problem of lack of standards has to do with the possible fragmentation of the market. If there are not enough major players interested in Cooperating Object technology, the remaining companies might not be strong enough to pull everything together.

The good news is that there are already some attempts to standardise ZigBee and UWB, which will probably play an important role as communication protocols that bring together networks of Cooperating Objects. However, this is just the beginning and further actions need to be taken in order for Cooperating Object technology to take off.

8.3.3 Confidence in Technology

Scientists and early adopters are definitely willing to make use of Cooperating Objects since they see an improvement in the way they solve certain problems. For example, the monitoring of animals or buildings represents an application domain where biologists and engineers have already started working together with computer scientists to deploy Cooperating Object technology.

For more sophisticated applications, currently available Cooperating Object technology is, for the most part, not able to deliver the desired characteristics, such as lifetime or robustness. Therefore, the immaturity of current solutions in individual fields hinders the adoption of Cooperating Objects in more general application domains.

However, in some cases this reluctance is based on prejudices, e.g. against wireless communication. It is hard to convince people that a new technology which is generally considered as more error-prone can deliver almost the same quality of service as the old, wired technology when designed carefully. As already mentioned in the section on business models, a trade-off between technology and costs has to be made for the most part.

8.3.4 Social Issues

Even if the technological issues are solved and the industry is able to pull together a set of standards that supports Cooperating Object technology, the end-users are still the ones that decide whether or not they will want to make use of this technology. For the most part, this implies that the market is "mature" enough, the price is acceptable and, more importantly, there are no social factors that hinder its acceptance.

The main question is whether or not the vast majority of people is willing to accept tiny computing devices "interfering" with their lives. People are not willing to have "big brother" watching them unless they see a benefit to it. The best example is the change in perception of security and privacy that happened in the U.S. as a result of the terrorist attacks of September 11. In order to increase their perception of safety, Americans have been willing to give up some of their privacy, something that would have had happened under other circumstances.

In general, people are reluctant to provide private information that might give an insight on their daily activities or habits and, therefore, if Cooperating Objects can be misused for this purpose, finding a killer-application might take longer than expected, if at all.

An increase in the awareness of security and privacy issues is surely needed for the proper understanding of the capabilities of Cooperating Objects, so that the end-users can put this new technology into perspective and not feel threatened. Legal actions will also help to define the proper use of Cooperating Objects.

8.4 Considerations for Research Programs within the EU

The set of predominant work areas discussed above has left the European Commission out of the loop, due to the special role they play in Europe. In this section, we expose a series of activities that might be of interest for inclusion in future research programs. It is obvious that our recommendations cannot propose implementation plans, since research frameworks and instruments are not part of our expertise. It would be the task of the interested reader to translate these recommendations into a workable plan within the possibilities of the EU.

We have divided our recommendations in three groups: specific research topics, general recommendations concerning research, and other recommendations.

8.4.1 Specific Research Topics

The gaps discussed in section 7.1 and the predominant work areas for research (section 8.1), indicate the need for research in specific topics that require an increased level of attention.

These research topics can be included in the corresponding descriptions of work of framework programs so as to encourage European researchers to submit project proposals for individual topics or combinations thereof.

The most prominent research topics include:

- The development of small, energy-saving and cheap hardware that can be used as part of individual nodes of a network of Cooperating Objects.
- Energy harvesting techniques that complement the use of batteries or even replace them.
- Data management algorithms that are able to cope with search, storage, aggregation of data, etc. in order to provide the basic functionality that makes up a network of Cooperating Objects.
- Research on non-functional properties such as quality of service, levels of reliability, etc. and their inclusion in the necessary algorithms that allows for adaptation.
- Work on the definition, design and development of adaptive system software and middleware that provide the level of support needed for heterogeneous application domains and networks.
- Security for Cooperating Objects in the form of algorithms that ensure certain capabilities of the network in order to make then suitable for more hostile environments.
- Research on social issues of Cooperating Objects in order to improve on the understanding of such systems both for research and for possible commercialisation.

For a more detailed description of these topics, refer to the rest of this document.

8.4.2 General Recommendations concerning Research

From the description above and the increasing complexity of such systems, it is obvious that there is a need for interdisciplinary research. For this purpose, the proposal of projects with partners from different backgrounds is highly recommended and, in some cases, absolutely necessary.

Another general consideration lies in the fact that most research involving Cooperating Objects should be performed in such a way that it is application-driven. The complexity and variety of application domains implies that the solutions proposed are only valid within those domains. Therefore, applied research should be encouraged, where possible.

Since Cooperating Objects are deeply embedded into the real world (whose complexity is hard to model and simulate), it is extremely important to build actual Cooperating Object systems and experiment in real-world settings in order to understand and solve the underlying scientific challenges of providing practical and efficient solutions to real world problems. The EU should hence actively encourage and support "systems research on Cooperating Objects", i.e., the scientific study, analysis, modeling, and engineering of effective software platforms for Cooperating Objects.

Finally, the integration of Cooperating Object technology in other systems is a general line of research that, in our view, should also be actively supported. This implies not only the integration of hardware, but also the increase in interoperability of software solutions with other research areas such as robotics, artificial intelligence, etc. that have traditionally used other techniques to perform research.

8.4.3 Other Recommendations

As for recommendations that do not really have to do with research, we have identified a series of work items in section 8.1 and 8.3. Most of these issues have to with legislative and standardisation actions that, to the extent possible, should be supported by the EU.

More specifically, the recommendations identified are the following:

- The support for public relations campaigns for Cooperating Objects would help in the increase of awareness of this technology that, when performed properly, helps actively in fighting against one of the major identified roadblocks.
- Similarly, there will be a need to support the creation of regulations and legislative actions that create a legal framework that supports the correct use of Cooperating Objects and hinders possible misuses of this technology.
- Supporting and encouraging the standardisation of hardware platforms and software platforms can help in the acceptance of Cooperating Object technology and can also be included in programs where industry participation is expected.

9 SUMMARY AND CONCLUSIONS

The field of Cooperating Objects is a very dynamic one, with such a high potential that it could really revolutionise our lives even in a deeper way than the Internet has done in these past years. In this research roadmap, we have tried to provide a thorough overview of the direction we expect research to take in the next 10 to 15 years and in the way researchers, industrial partners, end-users and financing institutions should work together to make Cooperating Objects happen in the near future.

For this purpose, we have given a thorough view of the state of the art regarding Cooperating Object technology, especially in the field of Wireless Sensor Networks, and have identified the trends that guide current research in the field. Knowing what is available allows us to have a glimpse as to what should be done next and what gaps and missing technologies should be addressed from the point of view of research in order to push this technology even further.

We have also given information about a probable timeline for the development of research and have attempted to characterise the points in time where major breakthroughs will allow for Cooperating Object technology to become mainstream. Additionally, we have tried to pinpoint major inhibitors and potential roadblocks and given concrete suggestions to avoid them.

Finally, we have provided a list of recommendations for future research and have tried to organise the proposed activities in such a way that the major players (the research community, the industry, the end-users, and the EU as financing institution) can collaborate and cooperate in such as way as to complement their efforts and pull in one direction.

To the researchers already working in this field, as it is our case, it is obvious that we are in the process of creating a technology that has the potential to change our lives for the better. However, we cannot work in this exciting endeavour alone. Only the coordinated actions from all interested parties will make the vision of Cooperating Objects a reality.

Feel free to jump in!

Bibliography

[1] H. Abrach, S. Bhatti, J. Carlson, H. Dai, J. Rose, A. Sheth, B. Shucker, J. Deng, and R. Han. MANTIS: system support for multimodAl NeTworks of in-situ sensors. In *Proceedings of the 2nd ACM international conference on Wireless sensor networks and applications*, pages 50–59, 2003.

[2] Mike Addlesee, Rupert Curwen, Steve Hodges, Joe Newman, Pete Steggles, Andy Ward, and Andy Hopper. Implementing a Sentient Computing System. *Computer*, 34(8):50–56, 2001.

[3] Jakob Bardram. Hospitals of the Future-Ubiquitous Computing Support for Medical Work in Hospitals. In *In Proceedings of UbiHealth 2003*. The 2nd International Workshop on Ubiquitous Computing for Pervasive Healthcare Applications, 2003.

[4] R. Barr, J. C. Bicket, D. S. Dantas, B. Du, T. W. D. Kim, B. Zhou, and E. G. Sirer. On the Need for System-Level Support for Ad Hoc and Sensor Networks. *Operating System Review*, 36(2):1–5, April 2002.

[5] P. Bonnet, J. E. Gehrke, and P. Seshadri. Querying the Physical World. *IEEE Journal of Selected Areas in Communications*, 7(5):10–15, October 2000.

[6] C. Borcea, D. Iyer, P. Kang, A. Saxena, and L. Iftode. Cooperative Computing for Distributed Embedded Systems. In *22nd International Conference on Distributed Computing Systems (ICDCS 2002)*, pages 227–236, July 2002.

[7] A. Boulis, C.C. Han, and M. B. Srivastava. Design and Implementation of a Framework for Programmable and Efficient Sensor Network. In *MobiSys 2003*, San Franscisco, USA, May 2003.

[8] CarTalk 2000. Web page: `http://www.cartalk2000.net`. Visited 2006-08-09.

[9] Haowen Chan, Adrian Perrig, and Dawn Song. Random key predistribution schemes for sensor networks. In *SP '03: Proceedings of the 2003 IEEE Symposium on Security and Privacy*, page 197, Washington, DC, USA, 2003. IEEE Computer Society.

[10] Cobis-collaborative business items. http://www.ctit.utwente.nl/research/projects/international/streps/cobis%.doc/. Visited 2006-08-09.

[11] COMETS. Web page: http://grvc.us.es/comets/. Visited 2006-??-??

[12] Embedded WiSeNts Consortium. Embedded WiSeNts Project Webpage. http://www.embedded-wisents.org/, 2006.

[13] CORTEX:. Analysis and Design of Application Scenarios. Deliverable D8, Lancaster University, May 2003.

[14] Paul Couderc and Michel Banâtre. Ambient computing applications: an experience with the SPREAD approach. In *36th Annual Hawaii International Conference on System Sciences (HICSS'03)*, Big Island, Hawaii, January 2003.

[15] CROMAT. Web page: http://grvc.us.es/cromat/. Visited 2006-??-??

[16] Cyberguide. Web page: http://www.cc.gatech.edu/fce/cyberguide. Visited 2006-08-09.

[17] A.K. Dey and G.D. Abowd. Toward a better understanding of context and context-awareness. Technical Report GIT-GVU-99-22, College of Computing, Georgia Institute of Technology, 1999. ftp://ftp.cc.gatech.edu/pub/gvu/tr/1999/99-22.pdf.

[18] Stefan Dulman, Tjerk Hofmeijer, and Paul Havinga. Wireless sensor networks dynamic runtime configuration. In *Proceedings of ProRISC 2004*, 2004.

[19] A. Dunkels, B. Grönvall, and T. Voigt. Contiki - a Lightweight and Flexible Operating System for Tiny Networked Sensors. In *First IEEE Workshop on Embedded Networked Sensors*, 2004.

[20] D. Estrin, R. Govindan, J. Heidemann, and S. Kumar. Next Century Challenges: Scalable Coordination in Sensor Networks. *International Conference on Mobile Computing and Networks (MobiCOM '99)*, August 1999.

[21] E. Fasolo, C. Prehofer, M Rossi, Q. Wei, J Widmer, A. Zanella, and M. Zorzi. Challenges and new approaches for efficient data gathering and dissemination in pervasive wireless networks. In *Integrated Internet Ad hoc and Sensor Networks*, Nice, France, May 30-31 2006.

[22] Christian Frank and Kay Römer. Algorithms for Generic Role Assignment in Wireless Sensor Networks. In *Proceedings of the 3rd ACM Conference on Embedded Networked Sensor Systems (SenSys)*, San Diego, CA, USA, November 2005. To appear.

[23] Frost & Sullivan. North American Commercial Vehicle Telematics Markets, July 2004.

[24] Frost & Sullivan. Remote Patient Monitoring Technologies Get Patients Wired for Health and Wellness, July 2004.

[25] Frost & Sullivan. North American RFID Markets for Automotive and Aerospace & Industrial Manufacturing, May 2006.

[26] Glacsweb. Web page: `http://www.envisense.org`. Visited 2006-08-09.

[27] C. Grosse, F. Finck, J. Kurz, and H. W. Reinhardt. Monitoring Techniques Based on Wireless AE Sensors for Large Structures in Civil Engineering. *EWGAE Symposium*, pages 843–856, 2004.

[28] D. Guerri, M. Lettere, and R. Fontanelli. Ambient Intelligence Overview: Vision, Evolution and Perspectives. 1st GoodFood AmI Workshop, Florance, July 2004.

[29] P. Havinga, P. Jansen, M. Lijding, and H. Scholten. Smart Surroundings. *Proceedings of the 5th Progress Symposium on Embedded Systems*, Oct 2004.

[30] W. B. Heinzelman, A. L. Murphy, H. S. Carvalho, and M. A. Perillo. Middleware to Support Sensor Network Applications. *IEEE Network*, pages 6–14, January 2004.

[31] J. Hill, R. Szewczyk, A. Woo, S. Hollar, D. Culler, and K. Pister. System Architecture Directions for Networked Sensors. In *Ninth International Conference on Architectural Support for Programming Languages and Operating Systems (ASPLOS'00)*, Cambridge (MA), USA, November 2000.

[32] R. Husler. Goodfood-Security. 1st GoodFood AmI Workshop, Florance, July 2004.

[33] ON World Inc. On world website. `http://www.onworld.com/`, 2006.

[34] R. H. in't Veld, M. Vollenbroek, and H. Hermens. Personal Stress Training System:From Scenarios to Requirements. Technical report, Roessingh Research and Development.

[35] M. Kero, P. Lindgren, and J. Nordlander. Timber as an RTOS for Small Embedded Devices. In *First Workshop on Real-World Wireless Sensor Networks*, Stockholm, Sweden, June 2005.

[36] M. Langheinrich, F. Mattern, K. Römer, and H. Vogt. First Steps Towards an Event-Based Infrastructure for Smart Things. In *Ubiquitous Computing Workshop (PACT 2000)*, Philadelphia, USA, October 2000.

[37] S. Li, S. H. Son, and J. A.Stankovic. Event Detection Services Using Data Service Middleware in Distributed Sensor Networks. In *IPSN 2003*, Palo Alto, USA, April 2003.

[38] Shuoqi Li, Ying Lin, Sang H. Son, John A. Stankovic, and Yuan Wei. Event Detection Services Using Data Service Middleware in Distributed Sensor Networks. *Telecommunication Systems*, 26(2-4):351–368, June 2004.

[39] T. Lindholm and F. Yellin. *The Java Virtual Machine Specification*. Addison-Wesley, second edition, 1999.

[40] Hongzhou Liu, Tom Roeder, Kevin Walsh, Rimon Barr, and Emin Gün Sirer. Design and implementation of a single system image operating system for ad hoc networks. In *MobiSys '05: Proceedings of the 3rd international conference on Mobile systems, applications, and services*, pages 149–162, New York, NY, USA, 2005. ACM Press.

[41] Ting Liu and Margaret Martonosi. Impala: A Middleware System for Managing Autonomic, Parallel Sensor Systems. In *ACM SIGPLAN Symposium on Principles and Practice of Parallel Programming (PPoPP'03)*, June 2003.

[42] S.R. Madden, M.J. Franklin, J.M. Hellerstein, and W. Hong. TAG: A Tiny Aggregation Service for Ad-Hoc Sensor Networks. In *OSDI*, Boston, USA, December 2002.

[43] M. Maroti, G. Simon, A. Ledeczi, and J. Sztipanovits. Shooter Localization in Urban Terrain. *IEEE Computer Magazine*, 37(8):60–61, August 2004.

[44] Pedro José Marrón, Andreas Lachenmann, Daniel Minder, Jörg Hähner, Robert Sauter, and Kurt Rothermel. TinyCubus: A Flexible and Adaptive Framework for Sensor Netwo rks. In Erdal Çayırcı Şebnem Baydere, and Paul Havinga, editors, *Proceedings of the 2nd European Workshop on Wireless Sensor Networks*, pages 278–289, Istanbul, Turkey, January 2005.

[45] Pedro José Marrón, Daniel Minder, Andreas Lachenmann, and Kurt Rothermel. TinyCubus: An Adaptive Cross-Layer Framework for Sensor Networks. *Information Technology*, 47(2):87–97, April 2005.

[46] Pedro José Marrón, Olga Saukh, Markus Krüger, and Christian Grosse. Sensor Network Issues in the Sustainable Bridges Project. *European Project Session of the 2nd European Workshop on Wireless Sensor Networks*, 2005.

[47] T. Mullen and M.P. Wellman. Some Issues In The Design of Market-Oriented Agents. In M. Wooldridge, J. Mueller, and M. Tambe (eds.), editors, *Intelligent Agents II: Agent Theories, Architectures, and Languages*. Springer-Verlag, 1996.

[48] Myheart. Web page: `http://www.hitech-projects.com/euprojects/myheart/`. Visited 2006-08-09.

[49] A. Ollero, S. Lacroix, L. Merino, J. Gancet, J. Wiklund, V. Remuss, I. Veiga Perez, L.G. Gutierrez, D. X. Viegas, M.A. Gonzalez Benitez, A. Mallet, R. Alami, R. Chatila, G. Hommel, F. J. Colmenero Lechuga, B. C. Arrue, J. Ferruz, J. Ramiro Martinez-de Dios, and F. Caballero. Multiple eyes in the skies – architecture and perception issues in the comets unmanned air vehicles project. *IEEE Robotics and Automation Magazine*, 12(2):46–57, June 2005.

[50] Onworld report on wireless sensor networks. Web page: `http://www.onworld.com/html/wirelesssensorset.htm`. Visited 2006-08-09.

[51] Oxygen. Web page: `http://www-robotics.usc.edu`. Visited 2006-08-09.

[52] Adrian Perrig, John Stankovic, and David Wagner. Security in wireless sensor networks. *Commun. ACM*, 47(6):53–57, 2004.

[53] G. P. Picco, A. L. Murphy, and G.-C. Roman. LIME: Linda Meets Mobility. In *21st International Conference on Software Engineering*, pages 368–377, Los Angeles, USA, May 1999.

[54] Everett M. Rogers. *Diffusion of Innovation.* The Free Press, New York, U.S.A., 4th edition edition, 1995.

[55] K. Römer and F. Mattern. The desing space of wireless sensor networks. *IEEE Wireless Communications*, 11(6):54–61, December 2004.

[56] Kay Römer. Programming Paradigms and Middleware for Sensor Networks. *GI/ITG Workshop on Sensor Networks*, pages 49–54, February 2004.

[57] Kay Römer, Christian Frank, Pedro José Marrón, and Christian Becker. Generic Role Assignment for Wireless Sensor Networks. In *11th ACM SIGOPS European Workshop*, pages 7–12, Leuven, Belgium, September 2004.

[58] S. Sahni and X. Xu. Algorithms for wireless sensor networks. *International Journal on Distributed Sensor Networks*, pages 35–56, 2004.

[59] Steven Shafer. The New Easyliving Project at Microsoft Research. *Proceedings of the 1998 DARPA/NIST Smart Spaces Workshop*, pages 127–130, July 1998.

[60] C. Sharp, S. Schaffert, A. Woo, N. Sastry, C. Karlof, S. Sastry, and D. Culler. Design and Implementation of A Sensor Network System for Vehicle Tracking and Autonomous Interception. In *Second European Workshop on Wireless Sensor Networks*, January – February 2005.

[61] Dana A. Shea. The Biowatch Program: Detection of Bioterrorism. Congressional Research Service Report RL 32152, Science and Technology Policy Resources, Science and Industry Division, November 2003.

[62] T. Sihavaran, G. Blair, A. Friday, M. Wu, H. D. Limon, P. Okanda, and C. F. Sorensen. Cooperating Sentient Vehicles for Next Generation Automobiles. In *Proceedings of the MobiSys 2004,1st ACM Workshop on Applications of Mobile Embedded Systems (WAMES 2004)*. 1st ACM Workshop on Applications of Mobile Embedded Systems, June 2004.

[63] Smart dust inventory control (sdic). Web page: `http://robotics.eecs.berkeley.edu/\~pister/SmartDust/`. Visited 2006-08-09.

[64] Michele W. Spitulnik. Design Principles for Ubiquitous Computing in Education. Technical report, Center for Innovative Learning Technologies, University of Berkeley.

[65] M. B. Srivastava, R. R. Muntz, and M. Potkonjak. Smart Kindergarten:Sensor-Based Wireless Networks for Smart Developmental Problem-Solving Environments. *Mobile Computing and Networks*, pages 132–138, 2001.

[66] P. Stanley-Marbell and L. Iftode. Scylla: A Smart Virtual Machine for Mobile Embedded Systems. In *3rd IEEE Workshop on Mobile Computing Systems and Applications (WMCSA 2000)*, Monterey, USA, December 2000.

[67] Sun RFID Industry Solution Architecture. Web page: `http://www.ascet.com`. Visited 2006-08-09.

[68] Arne Svenson. Executive Summary- The Safe Traffic Project, February 2005.

[69] J. Waldo. The Jini Architecture for Network-Centric Computing. *Communication ACM*, 42(7):76–82, 1999.

[70] Waternet. Web page: `http://www.rec.org/REC/programs/telematics/enwap/gallery/waternet.html`. Visited 2006-08-09.

[71] M. Welsh. Exposing Resource Tradeoffs in Region-Based Communication Abstractions for Sensor Networks. In *2nd ACM Workshop on Hot Topics in Networks (HotNets-II)*, San Josė, USA, November 2003.

[72] G. Werner-Allen, J. B. Johnson, M. Ruiz, J. Lees, and M. Welsh. Monitoring volcanic eruptions with a wireless sensor networks. In *Proceedings of Second European Workshop on Wireless Sensor Networks*, January – February 2005.

[73] K. Whitehouse, C. Sharp, D. Culler, and E. Brewer. Hood: A Neighborhood Abstraction for Sensor Networks. In *MobiSys 2004*, Boston, USA, June 2004.

[74] J. Yang, W. Yang, M. Denecke, and A. Waibel. Smart Sight:A Tourist Assistant System. In *ISWC'99*, 1999.

[75] Zebra RFID (Radio Frequency Identification) Product Tracking. Web page: `http://www.zebra.com`. Visited 2006-08-09.

Index

D

E

F

G

H

I

J

L

M

N

O

P

Q

R